Sexing the Parrot

Sexing the Parrot

Changing the World with DNA

WILSON WALL

CASSELL

First published in the United Kingdom in 1999 by
Cassell

Distributed in the United States by
Sterling Publishing Co., Inc.
387 Park Avenue South
New York NY 10016-8810

British Library Cataloguing-in-Publication Data
A catalogue record for this book is available from the British Library

ISBN 0-304-35221-7

Edited by Harriet Stewart-Jones
Designed by Geoff Green Book Design
Artwork by Richard Lucas

Printed and bound in Great Britain by MPG Books Ltd, Bodmin, Cornwall

Cassell
Illustrated Division
The Orion Publishing Group
Wellington House
125 Strand
London WC2R 0BB

Contents

Introduction

In writing this book, or more accurately, in thinking about this book as an idea, I was motivated by one simple thought: that knowledge is power. This century has seen many upheavals, political, social and scientific. While political changes may seem at first to be unconnected with the work of the laboratory scientist, it is important to realize that they are, in fact, intimately linked. Throughout history, our lives have been shaped by science, whether we knew it or not. When the Greek politicians first decided to use Greek fire in battle in the seventh century they were employing the power of chemistry. When the first recorded architect, Imhotep, designed the stepped pyramid at Sakkara in 3000 BC he used the science of geometry. And closer to home, the very fact that you are here and able to read this is a consequence of social changes resulting from advances in medicine in the early twentieth century. It was not so long ago that your chances of surviving childhood would have been small and infectious diseases would probably have killed you by now.

Things change quickly and the twentieth century has thrown up many advances which were beyond comprehension not so very long ago. Such things as travelling faster than sound, both in the air and on the ground, computer technology, colour television, all of these wonders have both added and detracted to the sum total of humanity. One of the areas of detraction in which everyone would agree is environmental pollution. Less often recognized is that the technology for these improvements in life style and associated problems of pollution has also been used in a different guise to clear up some of the very mess it caused in the first place. Modern technology has also helped to clear up problems arising from earlier, more primitive technologies. But perhaps the greatest advances have been those in the biological sciences, and in particular in genetics.

Of all the sciences – chemistry, physics, biology and the mistress of them all, mathematics – it is biology which has come of age in the twentieth century. The other sciences were formed and matured in earlier centuries, but it was impossible for biology to develop until certain foundations were laid. It could be said that alchemy powered the development of both physics and chemistry, although this is unlikely to be the whole story. By the same token, although witchcraft could have provided the impetus for the birth of biology it was, in fact, early attempts at medicine that encouraged empirical observations on the workings of life.

The study of biological systems is not always as easy as it may sound. For example, while everyone now knows that blood circulates around the body, it is incredibly difficult to demonstrate this without specialist equipment. This probably explains in part the late arrival of biology compared to the other sciences. While physical sciences are representable by straightforward formulas, biological systems are essentially complicated, dynamic, three dimensional structures. A cell is a dynamic system constructed from basic building blocks; the sum is greater than the parts and although cell chemistry follows the rules of thermodynamics, a cell is not a simple bag of molecules. This should be self-evident, but sometimes an element of reductionism can creep in where it is inappropriate.

Neither do we need to introduce a third and greater power, a supreme being, to help us understand how biology works. There is no reason for a teleological viewpoint on life. Just as the ancient questions – formulated in essence by the remarkable Anglican philosopher George Berkley – 'What is the sound of one hand clapping?' and 'If there is no one there to hear it, does a falling tree make a noise?' can be seen as irrelevant, so we can dismiss here any questions that rely on the existence of divine intervention. Nevertheless, the questions that we can answer are, in many ways, more amazing than these philosophical conundrums. When Bishop Berkley asked his questions he was pondering the world using a form of solipsism to determine whether there was even the existence of matter without consciousness to recognize it. It is not possible here to answer questions of this sort.

Of all the biological sciences it is genetics that has made the greatest advances and had the greatest impact over the past hundred years. This is not to play down the enormous strides and the immense importance of such fields as biochemistry and physiology. These are important, but as we will see it is becoming increasingly apparent everything to do with us, even our reaction to medicines, is affected by our genes. From the time when we first knew for certain that DNA makes up the genes in an organism to the present-day ability to manipulate those same genes, genetics has become pivotal. It may sometimes be called molecular biology, molecular genetics, cell genetics, or just plain genetics, but it is always essentially the same in nature: the study of the inheritance of characters.

It is not my intention in this book to produce a manual, nor do I attempt to explain in detail how techniques are carried out to produce a result. This book is much more about the result itself. If you know a gene sequence, what can you do with it and what are the implications for society? It is, after all, the practical applications of genetics that will affect us all over the next few years, from genetically modified foods to new and better medicines. Just as we do not need to know how a computer chip works to be able to exploit its potential, so we do not need to know how to clone an organism to realize the potential checks and balances that we, as a society, should consider before embracing this technology with open arms.

Genetics is, without doubt, one of the most powerful sciences to have come out of the twentieth century. Whether this power represents a threat to humanity or not depends entirely on how it is used. Some might say that it is physics, with the inherent ability to blow us all to pieces that poses the greatest threat, but that would be to ignore all the beneficial uses of physics. Just as medical uses of nuclear physics in X-rays and radiotherapy are the up side of physics, so the uses of genetics in the treatment of disease and the creation of productive crops could be of positive benefit to us all. But we cannot discount the potential downside, which is bioterrorism. As bioterrorism is seen as a real threat, and because politics and greed are very unpredictable motivators, it is suggested that although wild smallpox is extinct, it might be a good reason for keeping the two official remaining captive stocks of this virus alive, just in case we need to produce vaccines in the future.

Be in no doubt about this, genetics could help our world inordinately, but it is up to you where it goes. It is up to you as to what is acceptable; it is your society, not someone else's. If you see it as a problem then you are as much a part of the problem as part of the solution. On the other hand, do not make the mistake of jumping to what might be inappropriate conclusions. Confound Ralph Waldo Emerson when he said 'If a man sits down to think, he is immediately asked if he has a headache'. Risk the headache, sit down and think it through before making judgements. It is my hope that this book will help.

You may, at this point, be wondering about the title I have chosen for this work. Why parrots? What has DNA got to do with sexing them? And why on earth should we want to sex parrots anyway? There are good answers to all these questions. You'll have to read Chapter 6 to find out what they are!

ONE

The story so far

Searching for the material of the genetic code

It was in 1903 that the Wright brothers took off in the appropriately named 'Flyer' (sic) and achieved the dream of so many before them – the first manned flight in an aeroplane. Their longest flight lasted just less than a minute, yet so determined is the human spirit that less than a hundred years later commercial flights were carrying passengers faster than the speed of sound all the way across the Atlantic Ocean. In this same period of time, the science of genetics has risen from a descriptive natural history, surrounded by unknowns, into a science as rigorous and demanding as the traditional heavyweight disciplines of chemistry, physics and mathematics. The twentieth century is most certainly the century when biology came of age.

Those of us lucky enough to have seen this happen can only marvel at the determination of the scientists who worked on what had once been thought of as intractable problems in genetics and medicine. Since earliest times humans have wondered about inheritance of characters, of both themselves and their plants and animals. It had been known from the time of the first writing that some disorders were passed on in families from generation to generation, although how this happened was a mystery. Indeed, such familial disorders were often simply regarded as a family curse.

Just as the Wright brothers learnt from others who had worked on powered flying machines before them, so too did the scientists who took genetics from a rudimentary understanding that inheritance mattered, to the brink of gene therapy by the end of the twentieth century. Two giants of biology in particular laid the foundations for this work during the nineteenth century: the Austrian monk and biologist Gregor Mendel and the English naturalist Charles Darwin. During their own lifetimes it was not possible to put their work into a wider context, because knowledge of other areas of biology was lacking. When they expounded their theories they were little more than functional descriptions of their observations. This was a sort of natural history of heredity rather than a science of genetics. Only later did the fundamental nature of these ideas increase our understanding as more information was collected. But the principles which they described were exactly that – principles, and they have continued to guide our understanding of life and the manner in which it propagates itself.

The difficulties facing Darwin

Charles Darwin (1809–82) is associated in the minds of most people with evolution rather than genetics. But at a fundamental level evolution is intrinsically about the inheritance of characters and that is the science of genetics. During the period when Darwin was working there was no clear idea of what a mechanism of inheritance could be or how it could work. When *The Origin of Species* was published in 1859 it was difficult to understand exactly what the mechanism of evolution really acted upon, let alone what the forces of evolution were and how they exerted their influence. Although it had always been clear that any given species of animal only gave rise to animals the same as itself, there was no clear notion of what controlled this.

Darwin faced several major problems in the acceptance of his ideas. The first was the fact that the immensely long time scales required for evolution were far beyond the comprehension of most people in the nineteenth century. Even today it is hard to conceive of the aeons needed for evolution to work. Our sense of time scale is naturally based upon our lifespan; periods of more than a hundred years put events into the dim and distant past, even though they may be too recent to be measurable on a geological time scale. Estimates of the age of the earth vary, but it is usually given as about 4,500 million years, with life appearing about 3,500 million years ago. The oldest known fossils are of primitive bacteria from South Africa, dated at 3,100 million years old, but since bacteria are relatively complex organisms in their own right, it is reasonable to assume that life started some time before this. These vast figures make it very difficult to get an overall feel for the history of life on earth.

In the past, this difficulty has been compounded by the insistence of Biblical scholars that, based on a strict interpretation of the Bible, the earth was no more than a few thousand years old. Indeed in the seventeenth century, Archbishop James Ussher, in his *Annales Veteris et Novi Testamenti*, set the date of creation at precisely 4004 BC!

Another major problem faced by Darwin was that no clear, testable, hypothesis was available as to exactly how characters were inherited. For most people inheritance was a simple blending of parental characters in the offspring. This is not so surprising since those characters that are most easily observed are just those which are most difficult to quantify, like hair colour and skin colour in humans, or the colour of flowers in plants. To take an example of this confusion from the plant kingdom, if you cross a white snapdragon with a red one, the result is all pink flowers – the characters appear to have just blended together. It is only when these pink-flowered plants are cross-pollinated that the characters re-emerge. The first generation are still mainly pink-flowered but there are also some pure white and pure red flowers. This is a pointer to the fact that although the characters appear to blend, they still maintain their discrete nature and can be separated under suitable conditions.

Darwin attempted to address this problem of inheritance with the publication in 1867 of a two-volume work, entitled *The Variation of Animals and Plants under Domestication*. It was thought at the time to be a rather impenetrable work, so the original print run was quite small. What makes it so interesting today, however, is that it demonstrated that little had changed in the perceptions of inheritance since the time of Aristotle. It contained a series of complicated arguments regarding inheritance which were inevitably misdirected, because although cell theory was developing, there was a lack of any overall understanding of what was going on when cells grew and divided. This is hardly surprising given the very great difficulties involved in working out what was happening in a living organism from what was effectively a series of 'snapshots' of cells and tissues taken from preserved material.

In today's scientific world, the work of the Austro-British philosopher Karl Popper is taken to be the touchstone of science, the standard against which all experiments are held. Many scientists do not even realize that when they conduct an experiment to test a hypothesis they are following the work and theory of Popper. What he proposed was that scientific statements deny that something logically conceivable is actually achieved. Consequently it is just not enough for a statement to be counted as scientific simply because there are confirmatory observations supporting it. There must be a method, such as an experiment, which could disprove the statement by producing a result that it has said is impossible. This ostensibly convoluted idea is called the 'null hypothesis' and this is what all good science tests.

The pivotal point of a null hypothesis is that searching for a thing, anything, it could be fairies at the bottom of your garden or a high-energy particle in an accelerator, and not finding it does not mean that it does or does not exist. But saying it does not exist, which is essentially the null hypothesis, and then finding it, categorically disproves the original premise of non-existence. This, then, is the difference between natural history, which is essentially an activity of observation, and biological science, where experiments are conducted to test hypotheses.

Mendel and his peas

While Darwin was creating a synthesis of ideas based on observation, Gregor Mendel (1822–84) was describing a precise set of experimental results which would form the basis of modern genetics. This was the first time that any quantifiable idea of inheritance had been described accurately.

The son of a peasant farmer, Mendel was born in Brünn in Austro-Hungary, in what is now the Czech Republic. He entered an Augustinian monastery at the age of 21 and soon developed his considerable horticultural skills in the monastery gardens into research work on the way in which inheritance works. By 1868, when he was elected Abbot, his meticulous attention to detail had produced the results for which he is now justly famous.

Mendel's work on peas was quite remarkable. By carefully hand-pollinating the peas and systematically noting down the results he revealed an important clue to the mystery surrounding inheritance. He found that specific characters were inherited as discrete units, not as blended factors from the parents as had been believed previously. Mendel looked at many different qualities of his pea plants, such as whether they had round or wrinkled seeds, yellow or green cotyledons and long or short stems, all of which show a simple method of inheritance. Using these characteristics he followed their inheritance through the generations, observing the ratios in which they appeared, and finally came up with his concept of dominance.

We now know that all characters are controlled by genes and that genes come in pairs, only one of which is usually expressed. The gene that is expressed is called the dominant one and the one that is not expressed is called recessive. The precise mechanism controlling dominant and recessive expression is still poorly understood, but what we do know is that in any given species a dominant gene will always be dominant and a recessive gene will always be recessive.

Mendel must have been a most careful observer to have spotted the potential of peas from among all the other plant species which he had at his disposal. Or perhaps he was just lucky. Earlier he had tried using a species of hawkweed for the same experiments. Although these would have had the advantage of rapid growth and robust nature, hawkweeds use a form of asexual reproduction that would have made Mendel's breeding experiments impossible to reproduce. This process, called apomixis, means that the seeds can ripen without the need for fertilization. Consequently, a hawkweed plant with some particular characteristic will tend to pass this to all its progeny, as they are effectively clones of the parent plant. Since apomixis was not recognized at the time Mendel was working it was fortuitous that he did not persevere with breeding hawkweeds.

Peas, on the other hand, have a pollination mechanism ideally suited to breeding experiments. They usually undergo self-fertilization, and the pollen-producing anthers and pollen-receiving stigma are completely enclosed by the petals, which helps to prevent accidental pollination by insects. Not only that, but the anthers mature earlier than the stigma, so it is easy to remove the anthers from one flower and pollinate the stigma of a different flower.

When Gregor Mendel published the results of his research in 1866 they were greeted with complete silence on behalf of the scientific community. This was hardly surprising given that his work amounted to the creation of a completely new subject, the construction of a theoretical system that could explain so many observations, the laying down of the rules of inheritance, and a challenge to the wisdom of the day, all of which would have been good reasons for establishment rejection. But this was not the reason for the lack of acclaim. Mendel's results were published in German in a relatively little read journal called the *Verhandlungen des naturforschenden Vereines in Brünn*. Consequently they were simply overlooked.

It is a great misfortune that Mendel never received the tributes due to him in his lifetime, but all geneticists know of his work and the original explanation of what we now refer to as dominant and recessive characters. When it was republished in translation by the Royal Horticultural Society in 1900 it was finally realized just how fundamental this work was.

Microscopes and cell theory

With the publication of Mendel's work it almost immediately became possible to tie together several lines of research. In the 1830s two German biologists, Matthias Schleiden and Theodor Schwann, produced what we now call the cell theory. Although we now take for granted the idea that all living things are made up of cells, this has not always been the case. If you just consider for a moment the difficulty of seeing a cell and visualizing its activity, then it becomes easier to understand that until technical methods such as sample preparation and dyes became available, it was almost impossible for the early biologists to be sure exactly what they were looking at with their microscopes.

Three areas are crucial to microscopical work. These are the quality of the microscope, the dyes used to stain the samples and the method of sample preparation. With respect to the microscopes, there is little information regarding the instruments that early investigators used. What we do know is that the introduction by Zeiss in 1879 of the oil immersion lens, designed by a physicist called Ernst Abbe, allowed for far greater and more detailed microscopy. What Abbe managed to do was to take the construction of microscope lenses away from empirical methods and into the era of geometric optics.

Carl Zeiss was born in Weimar on 1 September 1816, the son of a toy shop proprietor. He became apprenticed to the instrument maker at the University of Thümingen in Jena, later working with instrument makers at Stuttgart and Vienna. In 1846 Zeiss opened his first workshop in Jena and a year later started to sell his own single-lens microscopes. That it had a single lens was probably due to two things; the first was that formal methods of microscope design did not exist and the second was the realization that a second lens also magnifies the imperfections of the first, so although magnification might be increased, so would aberrations. It was not long before Zeiss was producing the finest lenses in Germany, but he was always on the look-out for methods to improve them. At this time microscope manufacture was carried out by skilled craftsmen working essentially by trial and error.

Telescopes had been constructed using mathematical principles for many years and Zeiss was convinced that it should also be possible to do the same for microscopes. Leading mathematicians of the day tended towards the view that microscopes might just be too complicated to yield to geometric optics. At this point, Zeiss approached a 26-year-old physics lecturer with a simple question: Was it ever going to be possible to design a compound microscope mathematically? The lecturer, Ernst Abbe, quickly rose to the challenge and started work on his first designs in 1868. His first attempts turned out to be worse than those produced by traditional trial and error techniques. By 1872, however, Abbe had solved most of the earlier problems.

Soon Abbe realized that the glass used to make the lens was limiting performance. He thus enrolled the help of a chemist called Otto Schott. Schott was interested in developing new types of glass for optics to expand the range of glass produced by his family's glassworks. Abbe persuaded Schott to move to Jena and open a glassworks there. It is said that eventually it was possible for Abbe to specify the refractive index of glass, which he required, and for Schott to make it to that specification.

It was the commercial dynamism of Zeiss, mainly driven by a microscope called the No 1 'Jug Handle', which created and maintained their market lead. In 1884 a company based in London, Powell and Lealand, was already making an instrument that was capable of the ultimate resolution of the optical microscope. It should be remembered that the resolution of standard optical microscopes using transmitted light is limited by the wavelength of light, a purely physical factor, so it is possible to calculate the maximum resolving power of an optical microscope. The limit of resolution is the minimum distance between two structural elements, such as fine threads, at which they can be imaged as two separate items. Unfortunately, by 1910 Powell and Lealand had given up the production of microscopes under commercial pressure from Zeiss.

Advances in sample preparation

The nineteenth century was a period of intense development in microscopes. At the same time methods of sample preparation were also being developed in parallel to enable these new and better microscopes to be used to their fullest capacity. One of the pieces of equipment crucial to this expansion of microscopy and cell science was the automatic microtome. Before the development of this device, sections of plant stem for microscopy had been cut by sandwiching the material between two pieces of pith from the centre of an elderberry branch. This was then sliced using a botanical razor, which is nothing more than a cut-throat razor, hollow ground like a barber's razor on one side and flat ground on the other.

Microtomes are instruments used to cut extremely thin slices of plant or animal tissue that have been embedded in a rigid matrix. The embedding material was originally wax, which is still sometimes used today, although for ultra thin sections synthetic resins are now preferred. The material has to be sliced thinly because for microscopy it must be possible to shine a light through the sample. In the early days, cutting sections thin enough was, in itself, difficult because the sample had to be embedded in a material that would not simply fold and collapse upon the application of a blade. It had been possible to purchase an automatic microtome since about 1770, but it was the introduction in 1885 of the Cambridge rocking microtome, designed by Horace Darwin (son of Charles) of Cambridge Scientific Instruments, that made a significant step forward.

The fundamental difference between the rocking microtome and earlier machines was that it could produce a ribbon of contiguous serial sections. The first automatic microtomes produced by Cambridge Scientific Instruments between 1883 and 1884 were extraordinarily complicated and had a minimum step size of about 1/1000th of an inch. Horace's rocking microtome, however, was constructed using simple and high-quality engineering, which allowed serial sections down to 1/4000th of an inch (0.0006 mm) The machines were relatively cheap to produce and could cut thin serial sections of wax-embedded material of even and reliable quality. So robust were they that some have been known to be in use for more than 50 years. The ability to view serial sections enabled microscopists to build up a better idea of the interior of cells, both plant and animal.

The development of stains

Microscopes have three special qualities; these are magnification, resolution and contrast. Of these three, it is magnification that is least important. This may seem odd since the whole basis of microscopy seems, at first, to be magnification, but consider it further. The primary purpose of a microscope is to reveal detail. Without resolution and contrast, no amount of magnification will reveal anything in the specimen. Resolution is intrinsically tied up with physics, and the wavelength of light, so there is an ultimate resolution that a standard light microscope cannot go beyond. But compared to contrast, magnification and resolution are minor players indeed.

Contrast in any image is important; white on white or black on black will reveal nothing about the relationship between two objects. Whole organisms sometimes contain their own contrast, but when sliced up thinly enough for visualizing down a microscope nearly all biological material is transparent unless helped in some way. The normal method of increasing contrast is to stain the sample.

Early workers with microscopes used natural dyes such as saffron from crocus flowers (*Crocus sativa*) or cochineal, made from the dried bodies of a beetle, *Coccus cacti*, which lives naturally on several species of cactus in Mexico. Sheffield Art Gallery has a remarkable collection of microscope slides produced by M.L. Sorby which are stained with a variety of organic dyes, such as damson juice and port wine, all produced, with varying levels of success, in an attempt to increase contrast in the image. Of course, if the entire specimen were to be dyed uniformly, nothing would be gained because you would simply be swapping a colourless uniformity for a coloured uniformity. Biological material, however, is not uniform in density or composition, so usually a dye will only stain part of a sample. For example one dye may stain areas of protein, while another might only stain fat, resulting in a multicoloured image.

During the second half of the nineteenth century artificial dyes became available when 18-year-old William Perkin, later Sir William Perkin, accidentally created a mauve dye while trying to synthesize the anti-malarial drug, quinine. This new dye was greeted with great enthusiasm and was originally used to dye silk, but rapidly came to be used for all manner of things, even a postage stamp. Manufacture started in a shed at the bottom of his garden, but he soon moved on, ultimately building a factory for the very purpose of making dyestuffs. So successful was his manufacture of dyes that he was able to retire at the age of 36 and could devote the rest of his life to chemical research.

Although microscopists quickly learnt to use these new dyes to stain specimens, this was not their original purpose; most were produced for the textile industry as the new science of dye chemistry enormously increased the range of colours available to dyers, primarily for clothes. Thus in 1876 Heinrich Caro produced methylene blue and soon after methylene green, then 1878 saw the first production of malachite green and in 1884 Congo red appeared. Designed originally for dyeing cloth, all these dyes were made to an exact colour rather than an exact chemistry. In 1892 the chemical company Merck produced a list containing 51 artificial dyes, but at the close of the twentieth century there are approximately 3,500 dyes with a commercial value. Only a few of these are used in microscopy.

With every dye manufacturer using their own naming conventions and often with varying chemical compositions, it was soon realized that standardization was important. For example, there is a microscopical stain generally called orcein, which also travels under various different names such as archil, crottle, cud bear, orchil, orseille, persis and French purple; only the last of these gives a clue as to the colour, which is more reddish than purple. In 1922 the first method of classifying dyes was instituted by the Society of Dyers and Colourists in Bradford, England. This was the Colour Index, which resulted in dyes, for use both in textiles and in microscopy, which were reliable, gave repeatable results, and were defined by their chemical composition. Every dye now has a five-figure number, so no matter what the label calls it, if the number is there you can be sure of the chemical composition and result when it is used.

Cell theory and the discovery of chromosomes

Although cell theory had been anticipated by some scientists during the nineteenth century, it was not appreciated just how significant the cell was. In the second half of the nineteenth century, great strides were made in cell biology. The term 'genetic' had been in use since 1831, but it was the English scientist William Bateson (1861–1926) who defined the word 'genetics' for use in the way we know it today. Although he was a very astute scientist, Bateson never accepted the concept of natural selection. Neither did he accept the idea of chromosomes as genetic units.

At this time it was not entirely clear what the significance of some of the discoveries in cell biology and genetics would be. Since their research was so often pursued for its own sake, cell biologists felt no need to justify themselves, little realizing the profound nature of their work when combined with the knowledge of genetics which would be developed later.

As far as genetics is concerned, one of the most interesting bodies of work in cell biology during this period was that of Walter Flemming (1843–1905). Flemming was the first person to record having seen chromosomes, the carriers of genes in the cell, using his advanced techniques of fixing and staining cells. Of course at this time contemporary language was not equipped to describe all the newly observed parts of cells. Just as during this century language had to develop to handle the intricacies of computer science, so it was that in the latter years of the nineteenth century many new words about genetics were introduced. One such is the term 'chromosome', first coined by William Waldeyer-Hartz in 1888.

At the beginning of the twentieth century two major questions were being asked about chromosomes. The first was whether the number of chromosomes was constant. Did all the cells and tissues of an individual have the same number of chromosomes and did this number change during the normal activities of cell growth and division? It must be remembered that for these early researchers the chromosomes were effectively invisible for most of a cell's life. This is because, as we now know, while a cell is active the chromosomes are uncoiled to allow access to the necessary genes and it is quite impossible to tell where one starts and another ends. Any attempt to count them in this state is hopeless. It is only during a short period in the middle of cell division, called mitosis, that the chromosomes are tightly coiled enough to be recognized as individual entities. The second question people asked was, if the number is constant, then what is the number of chromosomes in a human cell? This question was to be hotly disputed for more than half a century until techniques of cell and tissue culture had developed enough for the answer to be found in 1956.

In 1902 Walter Sutton, an American, and Theodor Boveri, a German, proposed independently that chromosomes were important in heredity. While Boveri was originally using the eggs of a parasitic roundworm, he later moved on to sea urchin eggs. By surrounding the eggs with a mass of sperm it was occasionally possible to induce simultaneous fertilization with two sperm. What he demonstrated was that these eggs did not develop normally, so he made the significant conclusion that the correct number of chromosomes was essential for normal development. At the same time Sutton was also studying chromosomes, but using grasshopper testes. He correctly recognized that the number of chromosomes was constant, but also that there was a difference between some of them; there was an 'odd' chromosome left when all the others were divided up into pairs. In 1903 Sutton published *The Chromosome Theory of Heredity*. Another scientist, C.E. McClung, suggested that the 'odd' chromosome may be important in determination of sex, and in 1905 McClung also managed to show that in mammals females have two X chromosomes and males a single X and a single Y.

The choice of experimental material by these early geneticists might appear strange; after all roundworms and sea urchins are not always easy to come by. But such choices are often fortuitous for a number of reasons. For example with the sea urchin eggs double fertilization demonstrated that the correct number of chromosomes was needed for development. That would have been extremely difficult to show with most other organisms. The size of the individual cells can also be important. Much early research was carried out using salamander embryos, which have very large cells in the epidermis. So large, in fact, that detail can be seen that would be invisible in smaller cells. When investigating basic principles in cell biology and genetics, choosing the right experimental organism is half the battle. We rarely hear of the experiments using different material that failed for one reason or another and resulted in abandoned results, like Mendel and his hawkweed plants.

The chemistry of genes

In 1907, Thomas Hunt Morgan (1866–1945) firmly established that chromosomes were the basis of heredity by using another organism for his breeding experiments. This was the fruit fly called *Drosophila*. Again, it was a lucky choice of experimental animal. *Drosophila* has huge chromosomes, called polytene chromosomes, in its salivary glands. It also has the advantages that the adult fly is only a few millimetres long, it breeds prolifically and the two sexes can be quite easily distinguished with the aid of a magnifying glass.

Although by now chromosomes were known to be the units of heredity, it was still not known which chemical component of the chromosome was important. Neither was it known exactly how genes were organized or worked. In view of this, Sir Archibald Edward Garrod's research into inherited diseases in humans displayed remarkable insight. Garrod first published a paper in 1908 describing his research on several diseases caused by the incomplete metabolism of various dietary components. These diseases included alcaptonuria and albinism, both associated with problems in the metabolism of amino acids, and cystinuria, a kidney disease which leads to the excessive excretion of an amino acid called cystine. By studying families with these diseases he managed to show the first case of mendelian inheritance in humans. After many years of research in this field, Garrod came to a startling conclusion: that the cause of the disease was a defective gene affecting a metabolic pathway. It would be another 50 years before this impressive insight was shown to be correct.

Even with all this new knowledge of inheritance, the descriptions given to genes were still just functional explanations of observations; nobody knew what a gene was, what one actually looked like, or what it was made of. In 1911, the year that the first escalator was introduced to the public at Earls Court Underground station, Morgan produced a linkage map of his experimental fly, *Drosophila*. A linkage map is simply a representation of the relationship between different genes down the length of a chromosome. In actual fact, of course, at this time it represented the relationship between physical characters known to be caused by genes, rather than direct measurement of the genes themselves. In 1911 the linkage map only contained five genes, but by 1922 this map had been increased to show over 2,000 genes. It was this extensive work in genetics that earned Morgan a Nobel Prize in 1933. The word 'genome' was coined in 1920 by Hans Winkler to describe the total content of an organism's genes, but even with the increasing need for a specialized vocabulary to deal with genetics, exactly what a gene was made of remained unknown.

This was quite a remarkable situation. By the 1920s we knew that genes were on chromosomes, the way in which they were inherited, their action, the problems associated with defective genes, their susceptibility to damage by X-rays and in some species the order in which they were arranged on the chromosomes. But still no one could be sure what they were constructed of, or what they would look like when they were finally tracked down.

In the ensuing debate, the starting point was the knowledge that chromosomes were made of protein and DNA. This could easily be demonstrated by chemical analysis and by the way in which they stained for microscopy. Given that two major components could be identified it is hardly surprising that there should have been two camps with opposing views. The first suggested that genes were made of proteins while the other stuck to the assumption that genes were composed of a nucleic acid called deoxyribonucleic acid, or DNA.

Initially the reasoned logic suggested, incorrectly as it turned out, that the protein was the important part of the gene and that DNA was merely some sort of structural component. This assumption arose because it was realized that whatever it was that genes did, it was complicated, and therefore it would take a complicated molecule to achieve it. Proteins are chemically very complicated, whereas DNA is relatively simple, so DNA must be merely an adjunct to genes composed of protein. Although this line of reasoning seemed sound at the time, it turned out to be wholly inaccurate, but it took many years to demonstrate the error.

The next task was to identify the genetic material once and for all. This was not going to be easy, although there had been an important step in the right direction when in 1940 the experienced microbiologist E.L. Tatum recruited G.W. Beadle to work on the biochemistry of a pink bread mould. In 1946 Joshua Lederberg joined Tatum at Yale University to work on bacterial genetics. They demonstrated conclusively that one gene produced one enzyme. Until this time there had been assumptions, but no proof, that this was so. It had been suspected that since we are all so complicated, the analysis of single genes would be impossible because each gene produced more than one product. Although this 'one gene – one enzyme' hypothesis had been suggested many years before by Garrod, it was only now that proof was available. For this important body of work, Beadle, Tatum and Lederberg were all awarded a Nobel Prize in 1958.

In 1944 the *Journal of Experimental Medicine* published a paper by three Americans, O.T. Avery, C.M. Macleod and M. McCarty, who were also working on bacteria. This article described a series of experiments which finally proved that it was DNA and not protein that constituted the genetic material. It would not be true to say that genetics had been waiting for this information before being able to answer other fundamental questions, but it was true that this step was crucial in turning the functional descriptions of genes into an understanding of what genes really were. The reception to this important article was a little muted, but considering the date of publication, 1944, when the world was at war and many academics had given up pure research, this was not so surprising.

Genetics, disease and the double helix

In 1946 the first genetics clinic in the UK opened at Great Ormond Street Hospital for Children. This was a good choice as a site for such a clinic since it was recognized that for many genetic diseases and disorders it was crucial to diagnose the cause during a person's very early years. Indeed, if it was recognized that a child's problem was genetic, then suitable counselling could be given to the parents regarding both that child and future pregnancies. It was always hoped that alleviation of symptoms might be possible, even though no treatment that could be called a cure was available.

By this time there was a growing body of information as to the number and types of genetic diseases, but still little underlying detail as to what the metabolic problems might be. It was now necessary to explain how it could be that DNA, a chemically simple molecule, although of immense length, could successfully code for something as complicated as a whole plant or animal.

During the 1950s, as it became increasingly clear just how fundamental DNA was, extensive research was undertaken to try to determine the structure of this important molecule. It was finally in 1953 that the now famous double helix structure of DNA was elucidated by J.D. Watson and F.H. Crick, working in Cambridge. For this work they received the Nobel Prize in 1962, along with Maurice Wilkins who had helped produce some of the structural data of DNA using X-ray crystallography, while working with Rosalind Franklin at Kings College London. The Nobel Prize Committee is specifically precluded from awarding deceased scientists, so the untimely death of Rosalind Franklin stopped her from joining in the formal honours.

Chromosome numbers and the link between DNA and protein production

Another puzzle answered in the 1950s was the long-running debate as to exactly how many chromosomes were present in each human cell. This was finally resolved in 1956. Although it had been known for many years how many chromosomes other species, such as the fruit fly, have, the cells of humans had been rather more difficult to analyse. In the early 1920s, when material had been used from testicular biopsies, the consensus had been that humans had 48 chromosomes. It was not until new laboratory techniques became available that the true number of 46 was discovered by Jo Hin Tjio and Albert Levan in 1956.

In the field of medicine this represented another major step forward in understanding the links between certain specific inherited diseases and an individual's genetic makeup. It was suddenly realized that the presence of an abnormal number of chromosomes in a newborn baby was a significant cause of major genetic problems. Down's syndrome is undoubtedly the best-known example of this, but there are now known to be a considerable number of similar diseases and disorders. This knowledge resulted in the first prenatal diagnosis being successfully carried out in 1966.

But what was the relationship between a particular stretch of DNA and the protein molecules it encoded? By 1964 two scientists, the American geneticist Charles Yanofsky and the South African/British molecular biologist Sidney Brenner, had unravelled the answer to this question. Proteins are made up of strings of molecules called amino acids and DNA is made up of strings of bases. What Yanofsky and Brenner discovered was that the order of bases in a stretch of DNA corresponds to the order of amino acids in a particular protein. This was a significant step towards understanding protein synthesis and therefore the underlying nature of genes and disease.

In 1967 a singular observation by a group of American scientists suggested that some sections of DNA are associated with identical proteins in bacteria, toads and guinea pigs. This indicated that the genetic code was not only universal across all life forms, but even more amazingly, that some proteins are so important to the fundamental nature of life that they have not altered significantly through evolutionary time.

Other important advances in that year were the first reported synthesis of DNA in the laboratory using a single strand of viral DNA as a template and the discovery of the existence of 'stop codons', DNA codes that halt the production of protein at the end of a gene and thus control gene expression. As we shall see in Chapter 2, there are four different chemical bases in DNA and these are read in groups of three. By 1968 it was possible to determine that all of the 64 (4 x 4 x 4) possible codes had a meaning, and that they could be assigned to a function, either as a code for an amino acid, the building blocks of proteins, or as a stop signal. Clearly, as there are only 20 amino acids, each amino acid can have more than one codon to determine it.

Playing with DNA – progress in the late twentieth century

By 1979 it had been suggested that by cutting DNA into specific lengths using chemical analysis and enzymes it might be possible to find genetic 'markers' for specific diseases. This is an extraordinarily long and complicated process involving studies of entire families in which a particular genetic disease is found. If it is possible to find an association between a known sequence of DNA and the defective gene in a family, this is called a marker. Testing individuals for this marker will give some indication of the likelihood of that person carrying the disease gene. The absence of the marker makes it unlikely that the individual is carrying the defective gene.

In 1983 the first marker for a disease, muscular dystrophy, was located by Kay Davies and Robert Williamson, shortly followed by the identification of a marker for Huntington's chorea.

The early 1980s also saw the production of the first artificial chromosome in yeast. This gave researchers a tool of immense value over the forthcoming years. It enabled genes, either whole or in part, to be inserted into a yeast and then stably inherited from generation to generation, giving large amounts of DNA for research. Then in 1982 approval was given for the marketing of the first genetically engineered product, insulin. Until then insulin for diabetes had been extracted from pigs, but now a type of bacteria had been engineered with a copy of the human insulin gene. Although the product was of no use to the bacterial cell, when the gene was switched on the bacteria produced human insulin. This could be seen as the birth of genetic engineering as applied directly to mammals.

In 1984 Alec Jeffreys developed the technique to become known as DNA fingerprinting. This would have immense significance in forensic applications and matters of personal identity, specifically in paternity testing. By 1986 Louis Kunkel had determined the precise gene responsible for the disease called Duchenne muscular dystrophy, although it would still be many years before a method of prenatal diagnosis for this disorder would become available. The following year Kunkel described the protein product of the gene and called it dystrophin. This protein is vital to the structural integrity of muscle cells.

The invention of the polymerase chain reaction by American scientist Kary Mullis in 1993 had an immense impact on the development of genetics. This process, usually known as PCR, uses an enzyme to artificially replicate DNA, so that starting from only a single target molecule it becomes possible to generate enough DNA to analyse chemically. The use of PCR enabled all manner of diagnostic tests to be developed, as well as newer and more efficient methods of personal identification. With the introduction of PCR, genetics and molecular biology took a quantum leap into the future, so much so, in fact, that it is now unusual for articles in these areas not to include at least some work involving PCR.

It is inevitable in science that for every great discovery more questions than answers will be the result. The history of genetics has taken place in little more than one hundred years, but will, I am sure, continue to fascinate and amaze us all for many years to come.

TWO

The stuff of life

What is DNA and how does it work?

Deoxyribonucleic acid, better known as DNA, is fundamental to life. We now know this to be true on this planet at least, although it is unlikely to be true elsewhere in the universe. Even viruses that use modified forms of nucleic acid can only exist as parasites on organisms whose genetic material is DNA. Nucleic acid, then, in one shape or form, is the molecule that contains all the information necessary for the construction of all animals and plants, in fact for all life on earth. So far as we know there is no exception to this.

DNA is, if you like, a digital information code for all life. It should not be construed that as a digital system it is also binary in nature. Far from it, DNA, by its very construction, has an enormous flexibility and ability to absorb errors without detrimental effects. And although there are other forms of nucleic acid found in cells, it is the consistency of the fundamental nature of DNA that is so remarkable. This consistency of basic chemistry is one reason that we think that life is an evolved, indeed evolving, and uniform process. But it is only chemistry after all, and the story of how that simple chemical molecule, DNA, can create something as unique as you or me and the wide diversity of all the plants and animals on this planet is quite extraordinary.

Perhaps the strangest thing about all this, however, is that simply knowing about the chemistry of DNA does not completely explain why we are as we are. Nevertheless, if we did not know the chemistry of DNA we would be unable to understand anything at all about the methods by which we inherit the processes controlled by our genes, from the production of the smallest enzyme to the construction of something as complicated as our brains. Although plant and animal breeders were producing new and exotic strains long before it was known that DNA was the genetic material, today's detailed knowledge of the chemistry of DNA opens up extensive possibilities for the manipulation of DNA in the treatment of disease and the development of agriculture.

The chemistry of DNA

So what is DNA? In simple terms DNA is a molecule made up of just five elements: carbon, hydrogen, oxygen, nitrogen and phosphorus. These are the elements used to form the nucleotides, the simple chemical building blocks that are stacked up to form DNA.

A nucleotide is made up of three components: a base, a sugar and a phosphate linker. The most important of these in genetic terms is the base. There are four bases, called adenine, thymine, guanine and cytosine. As can be seen in the illustration on page 33, these all have similar though slightly different chemical structures. The other components of a nucleotide – the sugar and the phosphate – are always the same. The base is attached to the sugar via one of the nitrogen atoms in the base. This combined pair of base and sugar is now called a nucleoside. Once these are linked together by phosphate groups, the individual units are known as nucleotides.

When it comes to dealing with complex manipulations in the laboratory it is essential to understand not only the chemistry but the physics of this molecule, so here we will touch on the basic structure of DNA in slightly more detail. This is to give a flavour of just how elegant it is that such a simple molecule can generate such diversity of life. The genetic code is contained in only the four chemical bases: adenine, thymine, guanine and cytosine. A single strand of DNA is made up of a string of these four bases in any order. However, a single strand tends to form loops and has no rigid three-dimensional structure. It also has no mechanism by which it can

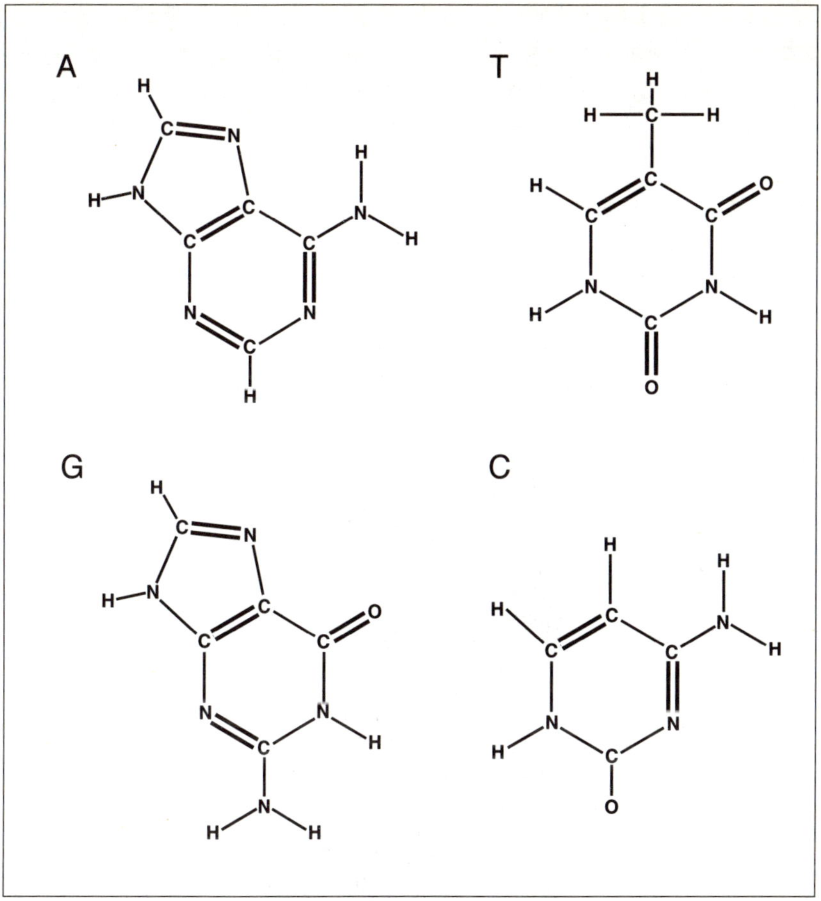

The four bases. The bases from which DNA is made are called adenine (A), thymine (T), guanine (G) and cytosine (C). As you can see, adenine and guanine have two rings; thymine and cytosine have just a single ring. This means that when adenine/thymine and guanine/cytosine nucleotides link up in DNA, the pairs they form are the same molecular size.

replicate itself. What it needs is a complementary strand of bases to which it can attach, forming a double-stranded structure. Each strand can then act as a template for the other, allowing accurate replication of both strands and thus replication of the entire double strand, which can then wind up into a double helix.

Bases do not differ widely in structure but there are two different types separated by details of their chemistry. These two types are known as purines and pyrimidines. The two purines are adenine and guanine; they are made up of two rings, one ring containing four carbon and two nitrogen atoms, while the other has three carbons and two nitrogen atoms. The two rings share two carbon atoms. The pyrimidines – thymine and cytosine – are much smaller, comprising single rings of four carbon and two nitrogen atoms.

The key to the stability of the double strand of DNA is that adenine can only link up stably with thymine, and guanine can only link up stably with cytosine. Now, each pair forms an almost identically sized molecular pair, but when they are stacked on top of

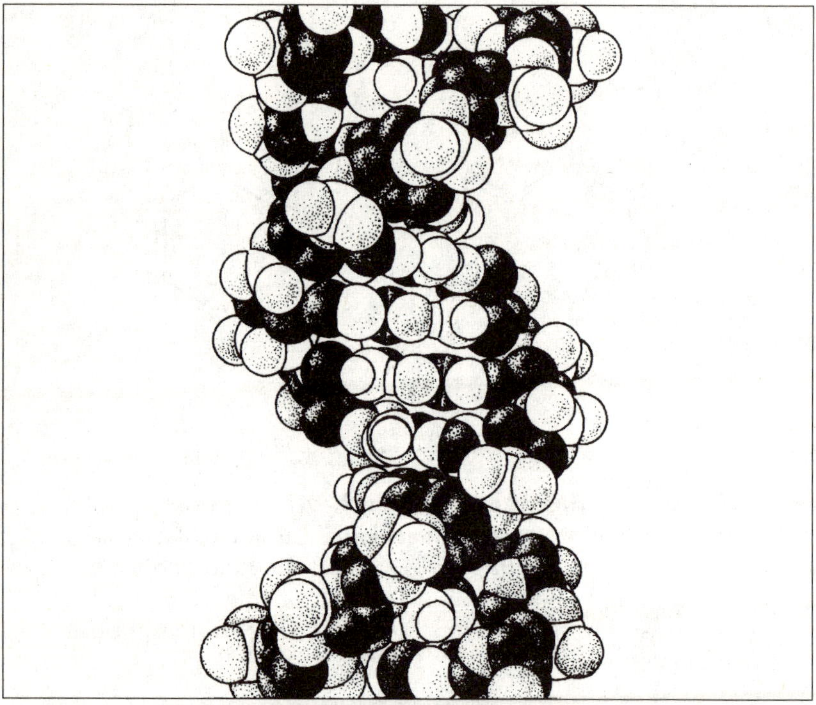

The DNA double helix. No molecule has ever been as recognizable as DNA. Is is used to advertise events and exhibitions all around the world. Is is a structure which can separate easily into two, producing a new strand of DNA as it goes. New base-pairs are added to the side branches as the strands separate. There are three billion base-pairs in the DNA of a person.

each other in any order each pair has to be linked to both the one above and the one below. This process of up and down chemical bonding results in the twisting of what might otherwise be a straightforward chemical ladder. This twist results in the famous double helix model of DNA as shown on page 34. This description of DNA holds true for all plants and animals. There are some exceptions where different bases are used, but these are generally confined to bacteria and viruses and even here they retain a remarkable similarity to the chemistry which guides our own inheritance.

The uniformity of genetic code across all organisms, from the smallest to the largest, from the simplest to the most complex, is perhaps one of the most surprising and fascinating things in biology. After thousands of millions of years the basic mechanism by which characters are handed down from generation to generation does not seem to have altered very much, if at all. Besides implying that life had a common origin, this uniform pattern of inheritance means we can work out fundamental principles of inheritance in one organism and extrapolate the results to others. The details may change, but the underlying principles do not.

The search for the double helix

It was two remarkable scientists, James Watson and Francis Crick, who finally found the solution to the question that had been puzzling scientists for many years when they worked out how DNA was constructed from its basic chemical parts. In the 1950s the chemistry of the constituent parts of DNA was understood, but not how they fitted together. For many years it had been recognized as crucial to know how DNA existed in cells, if we were ever to find out how heredity really worked.

Because this was seen as a big question in science, there were several groups working on the problem around the world. But it was in the Cavendish laboratory of the Department of Physics in Cambridge that Watson and Crick finally worked out the solution. Their analysis of DNA was carried out using a technique called X-ray diffraction. This technique produces two-dimensional photographs of the three-dimensional crystal structure of a molecule. It requires a very precise mathematical interpretation but this can be followed by model building, using beads and sticks, quite literally to construct a model on a bench top to see if all the angles and structures generated from the X-ray analysis could possibly fit together.

In the case of DNA this was not as easy a task as it might at first appear. As you can imagine, it would be impossible for an alien, for example, to generate a clear image of what an elephant is really like from a simple photograph. Two-dimensional images only ever give a flat representation of an object, even with our perception of perspective. So in the alien world of molecules, trying to image the structure of a molecule from a crystal, as happened with the analysis of DNA, where even the orientation was not known, was an immense problem.

X-ray diffraction crystallography was a technique developed by Sir William Lawrence Bragg, professor of physics at Cambridge, and his father, Sir William Henry Bragg. It takes advantage of internal reflections and refractions within the crystal, but the scale is so small that only something with a wavelength as short as X-rays can be used. It is a fundamental truth of physics that it is not possible to 'see' any object smaller than the wavelength that is being used to view the object. In X-ray diffraction, a beam of X-rays is fired at a purified crystal of the molecule under study. Because crystals have regular internal structures, they alter the beam in a regular and repeatable way. When the X-rays hit a photographic plate behind the crystal the resultant image represents a two-dimensional interpretation of the original crystal. The trick is to be able to analyse the picture to tell you what the target molecule actually looked like, and with intricate organic molecules this is quite a task.

When the Watson–Crick double helix model was produced, the biological implications were immediately apparent. The observation that adenine always paired with thymine and guanine always paired with cytosine made it clear that accurate self-replication of the DNA molecule could easily be achieved. It was a chemical fact, and nothing to do with the biology of the molecule, that if there was a thymine on one strand, then directly opposite on the other strand of the structure there would be an adenine. When it comes to replication, unzipping the double helix down the middle results in two halves, both of which can be replicated accurately. This is achieved using one set of enzymes that split the DNA double helix down the middle and another set that replaces the missing half of each DNA sequence accurately. Thus when a cell divides, each new cell receives a complete double helix of genetic material identical to that in the original cell.

Coding and control

After the detailed structure of DNA had been elucidated, one of the first questions which was asked was how on earth could a molecule like this, both simple and repetitive, be able to code for an entire organism? That is every single organism, not most, but all. Each species is different and each individual within that species, with some exceptions like identical twins, is unique. Yet each one is somehow described by the sequence of four chemical bases attached to a sugar backbone. This really did seem magical at the time and even as more and more becomes known of the way in which DNA works it still fascinates.

Almost immediately scientists realized that this extraordinary molecule could do the two things which were essential for life. The first was to self-replicate using only information held in the molecule itself. It had, after all, to code and control the enzymes of its own replication. The second was to control growth and development of both cells and organisms using only information held within itself. So our chemical of life, DNA, has to be able to control not only the development of the organism, but also accurate replication of itself through each cycle of cell division. This is really quite extraordinary, a molecule with the ability to exercise self-discipline! The search was on to understand the mechanisms that controlled DNA in the cell.

It is occasionally suggested that in some way a guiding force exercises power over the activities of the universe and all the living things in it. Such an idea does not necessarily help in our understanding of cellular biochemistry. Although it is true that biologists sometimes refer to cells and DNA as though they are in some way directed or intelligent, this is just a convenient shorthand. Nevertheless, it is possible to explain the workings of cells and the maintenance of heredity without involving ideas of teleology. In fact in many ways it is far more astonishing that the cell is a self-controlled system, with no intrinsic intelligence, from single, autonomous cells up to enormous multicellular colonies forming a dinosaur, a whale or a tree.

Even during the time that Darwin and Mendel were working, when only the vaguest ideas and concepts of cellular biochemistry were being worked out, it was debated among chemists as to whether cells followed the same rules of physics and chemistry as non-living systems. Just as the possibilities of natural selection were open to debate at this time, so too was the possibility that biological systems did not have to follow the simple rules of chemistry that controlled things such as the Bessemer Converter used in steel production, or the manufacture of soap. During the nineteenth century some people even wondered whether the laws of thermodynamics were necessarily applicable to biological systems at all, humanity being considered above and beyond the physical constraints of the universe. But today we know that, although complicated, biology follows exactly the same rules of chemistry and physics as the rest of the universe.

Theories of heredity

As we have seen, the very nature of DNA means it is able to replicate itself, and with very long sequences provide a template for life and the passing of characters from one generation to another, but this was not always known. Before the advent of Gregor Mendel and his development of clear ideas about inheritance there were numerous, mistaken theories of heredity.

One such was the theory put forward by Jean Lamarck in the early nineteenth century and later developed by a Russian, Trofim Denisovich Lysenko, to the great detriment of agriculture during the first half of the twentieth century in what was then the USSR. Lamarck proposed that characters acquired by an individual could be passed on to the next generation. For example, the neck of the giraffe was elongated over generations by having to stretch up to higher and higher branches for food. These notions, in many ways, were to be later confounded by ideas of both evolution and ecology. It was necessary to put the giraffe's neck into a unifying framework of genetics. The neck does not get longer through stretching, it gets longer because, if you like, the longer it gets the greater the advantage that species has over others. The length is controlled genetically, so it can be looked at as a series of mutations which increase the neck length under ecological pressure. If you cannot reach your food you die, but if your sibling can reach higher up the tree it will survive longer in adverse conditions, perhaps to breed and pass on this advantage. This is the essence of what Darwin suggested, not that stretching results in a long neck, but that the nebulous pressure of evolution will result in a long neck.

Lamarck's theories were developed by Trofim Lysenko to suggest that the heredity of crops could be altered simply by good husbandry. All gardeners know that good practice gives better results, but here was a man who said that if you grow a good crop in drought conditions, the ability to grow in drought will be passed on. This was not because of selective breeding, it was because the plant 'learned' what was required and passed it on to succeeding generations. He even went so far as to declare the work of Mendel incorrect. As director of the Institute of Genetics of the Soviet Academy of Sciences between 1940 and 1965, Lysenko had enormous influence on attitudes to genetics in the former Soviet Union. As a consequence he was, unfortunately, responsible for gravely hampering agricultural development in the whole of the USSR.

There is another hangover from this line of thought that plants and animals 'learn' characters that can be passed on. This is the idea that some sort of external power or force directs life in some way towards an ultimate goal. This was interpreted in the concept of a tree of life. The tree of life, as projected by darwinism, combined with remnants of church dogma, was first produced by Ernst Haeckel towards the end of the nineteenth century. Haeckel was professor of zoology at Jena in Germany. His concept of a genealogical tree of life had 'primitive' animals at the bottom and 'advanced' animals at the top. This resulted in monkeys, apes, humans, occupying the top of the tree, implying that humanity represented the greatest evolutionary advancement and perfection. Pictures of this type of tree often appeared in publications at this time, reproduced graphically as a diagram that started at the bottom of the page with bacteria and ended at the top with humans, with various branches along the way illustrating the diversity of life. A tree of life was always going to be a bad metaphor, especially if it was believed implicitly. Well, unfortunately for many years it was believed.

A bizarre misinterpretation of evolution arises at this point if you, the reader, think that you are at the top of some non-existent evolutionary tree. Evolution is not some sort of ladder to be climbed, it is the essence of adaptation to a specific environment. The concept of evolutionary superiority is incorrect, just as the idea of one species being more advanced than another is. We tend to mistake the evolutionary word 'advanced' with the passage of time, especially when looking at the fossil record. Consider the case of a desert-dwelling rodent. In terms of ability to survive in desert conditions it might be said that the rodent is more advanced than us humans. There is a flaw in this argument. It cannot be said that just because a creature is adapted to an environment it is 'advanced'. The idea of a family tree moving inexorably upwards to a pinnacle of evolution is fundamentally incorrect. It seems to be a leftover from religious ideas of humanity as the finest work of God. The best we can get from a 'tree of life' is some idea of the temporal relationship of origins and diversity between different species, not their 'higher' or 'lower' status.

The basics of life

Although it is the very simple chemical structure of DNA that makes it so reliable in its action, we should not assume that all we have in our cells is a nucleus full of 'heredity'. The story of inheritance is far more intricate than that. There are three basic sites at which DNA occurs in a cell. The biggest, and most important, of these three is the nucleus; this is where most of the genes that control the overall structure and function of the cell are located. The other two sites of DNA are structures called mitochondria and chloroplasts. The mitochondria, the powerhouses of the cell, are essential for work. This is where the sugars are broken down to release energy for growth and life. Without mitochondria there is no energy; and without energy there is no life. Chloroplasts are normally only found in plants. These are the structures inside the cell responsible for capturing the energy of sunlight and using it to convert carbon dioxide and water into sugars.

The process of photosynthesis – this conversion of water and carbon dioxide into sugars using the energy of sunlight – is a quite remarkable piece of chemistry. The plant manages to take in carbon dioxide from the atmosphere for use in the reaction, even though it is present in air in only tiny amounts. The composition of air under standard conditions is about 78% nitrogen and 20% oxygen. The remaining 2% is made up of many other gases, and carbon dioxide comprises only 0.03%, although this can vary slightly from place to place.

Although we tend to think of carbon dioxide as a 'greenhouse' gas and as a threat today, it is interesting to note that the levels in the atmosphere now are much lower than they were 345 million years ago, before the Carboniferous period. Of course the difference is that at that time the extant organisms were adapted to life in that atmosphere. We would be unlikely to survive in such conditions. Life, at any time, is very delicate; it is only when change takes place over extraordinarily long periods that adaption and evolution makes survival possible. It was during the Carboniferous that the coal measures were laid down, locking up enormous quantities of carbon dioxide from the atmosphere and leaving us with today's very low atmospheric concentrations.

DNA in more detail

Whatever of type of DNA is looked at, the molecule follows a series of general rules of construction. As we have seen, DNA is made of four bases – adenine, guanine, cytosine and thymine. These are usually shortened to A, G, C and T respectively. The proportion of these four bases varies according to the organism being investigated, but the DNA from different cells from the same organism will always have the same composition. Likewise, DNA from closely related organisms have a similar composition, whereas distantly related species tend to have DNA with widely differing base compositions. This is, of course, exactly what we would expect. Since DNA controls the growth and development of an organism, different organisms should have different DNA. We can turn this around and use it to decide how closely related different species are by looking at the variation in DNA.

There is a rule of base equivalence associated with DNA which gives us a clue as to the structure of the molecule. This rule, sometimes called 'Chargaff's rule' after its discoverer, Erwin Chargaff, who studied DNA during the early 1950s, says that the number of thymine bases always equals the number of adenine bases. Similarly the number of cytosine bases always equals the number of guanine bases. But, the number of guanines and cytosines does not equal the number of thymines and adenines. We can put these observations together to say that the number of adenine and guanine bases always equals the number of thymine and cytosine bases (A + G = T + C). This information gave the early researchers a pointer towards the fact that the A and T bases fit together in some way and so do the G and C bases.

At this point it would be useful to look in more detail at the the chemical nature of DNA. As we have seen, the four bases are held together as a chain of nucleotides. Each nucleotide comprises a sugar, a phosphate group and a base. All this chemistry may seem far removed from useful applications of science, but without it we would not be able to even begin to use our knowledge for any practical purpose. When referring to a DNA sequence it is normal to ignore the sugar and phosphate linkage as these are common to all bases. Consequently we can write a DNA sequence as a series of letters of the four bases. For example,

AATCGTCGAATTGGTGCAGATCAGAT

We also know that the bases appear as pairs, with A linking to T and C to G. Thus the double strand could be written:

AATCGTCGAATTGGTGCAGATCAGAT
TTAGCAGCTTAACCACGTCTAGTCTA

The bases do in fact link together in a very ordered manner. They stack up on top of each other and at the same time are linked together sideways so that the overall structure is very stable. Simply by virtue of the chemistry, the DNA molecules assemble themselves as a right-handed double helix, similar to a spiral staircase. Each rotation of the helix takes approximately ten bases to complete, so although written out as a string of letters, any sequence will be twisted through 360 degrees every ten letters and the two strands of the structure cannot be separated without unwinding the helix. This structural stability is part of the reason that DNA survives so well in adverse conditions. It also means, of course, that when it comes to replicating the molecule a barrage of enzymes are required to unwind, replicate and put it back together again.

The link between any two bases in a pair is, in chemical terms, relatively weak. This means that it is possible to 'unzip' the DNA molecule down its length. These weak bonds are called 'hydrogen bonds' because they are formed by the weak attraction found between hydrogen atoms. They are easily broken by heating and some chemicals, and while this weakness does not affect the DNA sequence itself, it is a property used to great advantage in the manipulation of DNA in the laboratory.

In contrast, the bonds that hold the bases together down the length of the DNA molecule are usually immensely strong. These 'covalent bonds' require considerable energy to break them, and they make the DNA an extraordinarily robust molecule. It is covalent bonds which make chemicals such as sodium chloride, common salt, so very stable.

As we mentioned earlier, the physical sizes of the bases are such that the distance between the two strands remains the same down its entire length, like a ladder. This is because a pair made up of T and A is the same size as a pair made up of G and C, each pair comprising a two-ringed molecule and a one-ringed molecule.

Because of this pairing, the length of a DNA sequence is always expressed in terms of its number of base pairs, or bp. One thousand base pairs is a kilobase pair, denoted kb, and one million base pairs is a megabase pair, denoted Mb.

It may seem unnecessary to have units as large as a megabase pair, but as we shall see, life is controlled by huge amounts of DNA, and we need a short-hand notation for the enormous numbers of bases involved. These very large units also explain why sequencing even small gene sequences takes so long and why working out the complete sequence of DNA in the genetic code of humans will take large numbers of people many years to complete.

The Human Genome Mapping Project (HGMP) is a massive international project set up to determine the entire sequence of DNA in humans – the human genome. This is a vast and scientifically demanding undertaking on a scale with sending a man to the moon. When finished, the HGMP will give us a consensus sequence of the entire human genome. It is called a consensus because we know that there is sequence variation between individuals, and only certain highly conserved genes will be the same in different people. Overall the level of variation is surprisingly small, often being confined to sequences of DNA that do not seem to have any practical purpose. These areas are of of immense value in identifying people through DNA fingerprinting and in tracking disease genes through families.

What is a gene?

It is a curious fact that the amount of DNA in a cell does not necessarily correspond to the number of genes. If the genes are imagined to be beads on a string, then the spaces between the beads comprise what is sometimes called 'junk DNA', sections of repeating sequences that do not code for anything. The purpose of these sequences is not understood and the amount present varies enormously between species. For example, salamanders, which are newt-like amphibians, have between 5 and 20 times more DNA than humans. Most of this is so-called 'junk DNA'. In contrast, the muntjac deer has the equivalent of 0.7 times the amount of DNA of humans. This variation in DNA content has puzzled biologists ever since it was discovered. No one has been able to explain why all this additional DNA is found in some species, while others do not seem to need very much at all.

So how many genes does the human genome contain? An awful lot is the most simple, if slightly facetious, answer. Most of these genes are working away without any apparent effect. They may control our height, our hair colour, our facial features and, indeed, the way in which our metabolism works, but in day-to-day life they are effectively unseen. It is only when something goes wrong and a defect appears, such as those responsible for the genetic diseases muscular dystrophy and cystic fibrosis, that genes demonstrate their importance in the maintenance of our very complicated bodies. Thus the fate of most of our genes is to work silently in the depths of our biochemistry and physiology, their essential importance to life unsuspected until they fail and make their presence felt.

Before it is possible to decide how many genes there are, we also need a clear idea of exactly what we mean by the term 'gene'. It is, at the moment, quite difficult to decide what a gene is, because it is necessary to determine exactly where a gene starts and where it stops. This is called the 'reading frame' of a gene.

Broadly speaking, we can think of a gene as any section of DNA which is transcribed. Transcription is the process of reading the DNA and producing some sort of product from the information. For scientists this definition can also cause problems. We have to decide if this sequence should also include all the ancillary sections of DNA that control and switch on and off the gene activity. Previously, estimates of gene numbers in humans have ranged between 50,000 and 100,000. This very large range is really not very helpful, but one thing has always been generally agreed upon: half of all the genes in humans are associated, either directly or indirectly, with the construction and running of our central nervous system.

The human genome, so far as we can estimate, is made up of approximately 3,000 million base pairs (3,000Mb). To take this value and estimate gene numbers we have to estimate how common genes are along the DNA. As we have said, not all DNA is genetically active and the density of genes along a chromosome does vary. Estimates range from one gene every 40kb to one gene every 50kb. We can use this figure to narrow the range of our estimate to between 60,000 and 75,000 genes in a human being.

This is a genuinely huge number and it is a constant source of amazement that so many of us have so few disabilities associated with genetic problems. There is so much that could go wrong, it is surprising that so little does. This is a reflection of the remarkable ability of the genome to not only transcribe DNA accurately, but also to replicate itself repeatedly, not only within an individual, but more importantly across countless generations, with so few errors.

The DNA code

As described earlier, there are usually four bases making up DNA. I say 'usually' because exceptions do crop up; chemically modified bases sometimes occur in nature or are used in the laboratory in experiments to elucidate DNA structure and function. The information contained in the DNA is stored in the form of a code made up in triplets of bases called 'codons'. Remarkably, this triplet code is the same across the entire animal and plant kingdoms, giving us yet another demonstration of just how fundamental DNA is to the development of life on this planet.

So how is the information contained within the DNA triplet code used? In order to understand this, it is necessary first to look at the chemical structure of the substance produced – the proteins. Proteins are large, complex molecules made up of chains of units called amino acids. They are fundamental to life and far more complex in structure than DNA. Proteins regulate the activity of the cell in all species of life and are the most important structural part of animal cells. In fact, some tissues, such as muscle, are nearly all protein. Each protein gains its complexity from two things: the first is the amino acid sequence and the second is the way in which it is folded to make a three-dimensional structure. As we have mentioned earlier, there are about 20 different amino acids, each with its own name, and they come in a variety of sizes and shapes based on a configuration of carbon, nitrogen, oxygen, hydrogen and sometimes sulphur atoms. Some amino acids are in the form of carbon rings, some have rings of carbon and nitrogen and some are just straight molecules.

Although amino acids are not particularly large in themselves, they all share the ability to be joined together into very large molecules indeed. Proteins have no theoretical limit on their size; most contain more than 100 amino acids and some are a good deal larger than that. Once joined together in the correct sequence, the string of amino acids can be modified in the cell by folding it in specific ways to produce the final, functioning protein.

In basic terms, then, the sequence of bases in the DNA translates into a sequence of amino acids, and thus determines the type of protein produced. As there are four bases in groups of three, there are 4 x 4 x 4 possible combinations, 64 in all. However, since there are only 20 different amino acids there would appear to be 'spare codes'. It is known that some of these are taken up as signals to start and stop amino acid production but even this does not account for all the possible codes. This apparent redundancy in DNA coding potential comes about because of a built-in level of error acceptance called the 'wobble hypothesis'. Stated simply, a triplet of three bases has an ability to vary one base and yet still code for the same amino acid. So, for example, there are four different codes for the amino acid called alanine, four for glycine and four for valine. Each of these different codes for each of the different amino acids start with a common pair of bases, GC in the case of alanine and GG in glycine. The last base of the triplet can vary without affecting the amino acid produced. Arginine and leucine are exceptional, both having six different methods of coding, not all having the first two bases the same.

The amino acid code is one of the most amazing aspects of this whole process. Every function in the body can be expressed in terms of combinations of these 20 amino acids. The proteins formed by stringing amino acids together are of potentially enormous variety because the number of amino acids in a protein can be huge; any combination of the 20 amino acids in any numbers. Some proteins are quite small. Insulin, for example, which was the first protein to be completely sequenced, finally being completed in

1953, is made up of less than 100 amino acids. Others can be truly massive, like dystrophin, the protein that is defective in muscular dystrophy.

Proteins also change in size during their lifespan. Insulin is a protein of 51 amino acids when it is ejected from the cells of the pancreas. When it is originally made, however, it is actually 82 amino acids long. What happens is that after the gene has been transcribed from the DNA and converted into an amino acid sequence in the small subcellular organelles called ribosomes, other enzymes in the cell crosslink the two ends of the protein and snip out the intervening 31 amino acids from the pro-insulin to convert it into the active form.

In the early days of protein research it used to take a long time to work out the sequence of amino acids in a protein. For example, when the digestive enzyme called chymotrypsinogen was sequenced, it took the equivalent of 15 man-years to complete the task and to discover that it contained 246 amino acids. Today it does not take so much effort to sequence proteins, but it is still a long and difficult job. So why do we need to know the amino acid sequence of a protein? The answer is that information about the sequence can allow us to go one of two ways: we can either construct an artificial copy of the protein or we can track back to the gene as a first step on the road to gene therapy.

In the case of insulin it was soon realized that this hormone, which is so important in the control of blood sugar, is virtually identical in all mammals. There are minor differences between species, but they are so close that pig insulin can be used to control diabetes in humans. The medical possibilities opened up by this realization were immense.

Protein production

At this point it would be useful to look in a little more detail at how proteins are produced. All the enzymes and structural proteins of every organism are coded using this system. Other functions of the cell are controlled and coordinated by these proteins once they are made. A number of steps are involved in the production of a

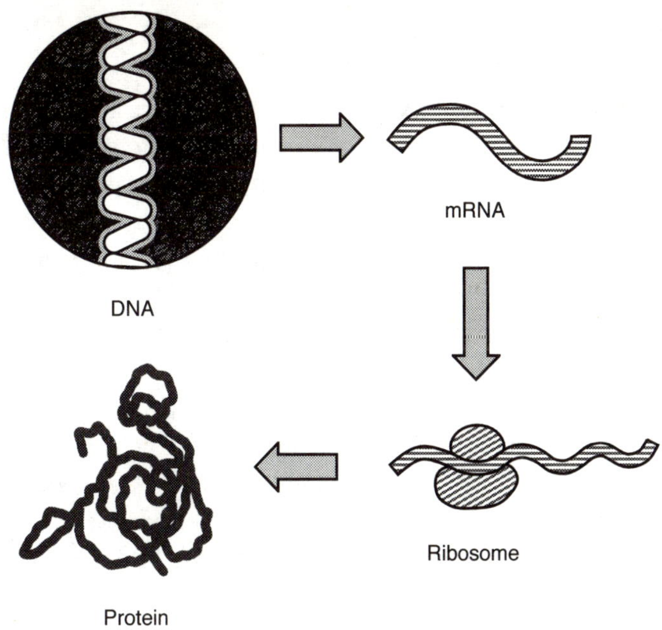

The route from DNA to protein. DNA is transcribed to form messenger RNA (mRNA) which travels out of the nucleus into the body of the cell. Here it comes into contact with a subcellular organelle called a ribosome. In the ribosome the mRNA is translated into a protein. This translation occurs by the addition of appropriate amino acids in the order dictated by the original triplet codes of the DNA.

protein, as shown in the illustration above. First, DNA has to be transcribed; that is, a gene is read, three bases at a time and a small, complementary sequence of bases produced. (The complementary sequence is slightly different in that instead of thymine it uses another base called uracil.)

This complementary sequence is attached to a much larger molecule called messenger RNA (mRNA). RNA (ribonucleic acid) is very similar to DNA, the difference being that in RNA the deoxyribose sugar is replaced by a ribose sugar. The mRNA is transported from the nucleus to a small body within the cell called a ribosome, where it attaches an amino acid of the correct type to an ever-growing string of amino acids. This process is called 'translation', the base triplet being translated into an amino acid.

So proteins are not read directly from DNA and produced in the nucleus; they are produced in the ribosomes following the processes of transcription and translation from DNA to mRNA to protein. Because mRNA is a complementary, or mirror image, sequence and DNA has two strands it is important that only one of these is used for production of mRNA.

It has been suggested recently that the way the code evolved minimizes errors during translation from DNA to protein. This is not necessarily the only explanation of the evolution of the code, but does have some persuasive elements. Clearly, if errors occurred in translating a gene into a protein it would be best if the error resulted in a very similar amino acid being incorporated into the protein, thereby minimizing the effect on the final protein. Just as we saw earlier that the last of the three bases in a triplet code can be altered without necessarily changing the amino acid produced, so if the code is changed for an amino acid, it is replaced by one which is chemically very similar. An alternative view would be that the present DNA code arose from a simpler version where similar amino acids were coded for by similar DNA. Whichever is true we can see that evolution shaped the code, shaping the very way that the chemistry of life is interpreted.

Mistakes are made

Although we have said that DNA gives a remarkably stable system of codes, it should not be assumed that errors never occur in transcription, translation and replication. Indeed it has been suggested that accumulation of errors during replication is part of the normal ageing process. What we do know for certain is that errors of replication and transcription are pivotal in the causes and development of many cancers. However, since mRNA transcription uses a fixed DNA template, occasional errors resulting in the wrong amino acid being incorporated into a protein will not be perpetuated. The vast majority of protein will therefore be correctly constructed and work properly, odd molecules being swamped by correct ones.

Studies of DNA replication show that errors occur far more often than one would at first imagine. If these were just left uncorrected, protein synthesis could become severely compromised so there must be some process for checking and correcting mistakes. The enzymes involved in DNA replication do have some ability to 'proofread' DNA as it is made, checking that nucleic acid bases have been correctly incorporated into the new strand during replication. This is done by comparing the match between the new strand and the old, template, strand. Errors should be recognized and put right before disaster strikes and the cell ceases to function correctly; unfortunately this does not always work.

There is also the built-in 'wobble' of the DNA code discussed earlier. Since there are 20 amino acids to code for with a code capable of 64 different permutations, it is easy to see that some mutations in DNA will result in a different base but the same amino acid. This wobble in sequence is associated with the last of the three bases in the code. A change in the first base of the triplet has a far more devastating effect then either the second, or least of all the third. Thus a mutation in DNA may not make any detectable difference in the amino acid sequence of a protein. But what about mutations that do change a sequence sufficiently to alter a protein, or that switch on a gene that should not be switched on? These are, after all, causes of genetic diseases and the starting point for cancers of all sorts.

Protection against disease

When we have a cold, caused by infection with some organism far smaller than we are, but able to reduce us to a shivering, sniffing, sneezing, wretched state, we wonder how on earth it is that such a tiny bug can wreak such havoc. Strange as it may seem, it is not the infectious organism that raises our temperature and makes us feel so bad, it is our own defence system. Our immune system causes our body to react in this way and although it usually succeeds in eradicating the invader in the end, it creates all sorts of problems for us in the mean time. So why should our immune system be so destructive?

One theory to explain this, worked out from simple logical deduction, is that our immune system did not originate simply to combat infectious organisms. Most likely its primary purpose was to protect the body against itself. Protection against invaders is a secondary benefit of immunology.

So why would it be necessary to protect yourself against yourself? Well, mutations can be very destructive if they are cancerous. So the question is, if so much importance is attached to our immune system eradicating potentially harmful mutations, how often do mutations occur in a large and long-lived organism like a human?

Taking a simple set of rounded figures we can demonstrate the idea of mutation protection. Bearing in mind that mutation rates vary considerably from organism to organism and, more importantly, from gene to gene, the mutation rate within a given gene might range from, say, one in 100,000 to one in 100,000,000 per cell division.

As can be appreciated, this is a very small error rate in DNA replication. But humans have about 100,000,000 cells per gram of tissue, so where are all the mutations going? Luckily, nearly all tumour precursor cells have a recognizably different cell surface to normal cells and it is the cell surface that your immunological system recognizes. So for most mutations the change in the cell surface means that the cell is recognizably different and is instantly seen as 'foreign' and therefore destroyed by one of the immune responses. And it is this which is seen as the primary use of the immune system in biological terms, althugh as far as most are concerned it is the eradication of microorganisms which is the important activity of the immune system.

Of course invading microorganisms also have different cell surfaces to human cells, so they are also recognized as foreign and destroyed in the same way. With the rapid reproductive ability of bacteria and viruses the system is very quickly overloaded, hence the extreme reaction of the body.

Filling your head with nonsense

It is often suggested that DNA is a digital code for all life, the commonest comparison being with digital computer storage. However, this is a rather simple view of a series of very complicated interactions between genetics and environment that result in a complete and well-adapted organism. So before we become obsessed with this digital analogy, remember that DNA is in reality a highly precise, yet flexible, system of transmitting data. The important difference between DNA and digital transmission is its flexibility.

Thus although there are good reasons for using a digital analogy it should not be pressed too far. If we start thinking about animals as being no more than sophisticated machines we are likely to strip them of their essential dignity. Simple observation of unpredictable animal behaviour will show how wrong this course of action would be.

Another computer analogy that had a temporary vogue in some quarters recently was the idea of an animal or human brain as some sort of super-sophisticated computer. This analogy can be dismissed as unrealistic because there is a difference between computer memory and biological memory which can be easily explained by a direct comparison of examples.

For a start, the way in which the human brain stores information is quite different from data storage in computers. Consider the quantities involved for a start: we know that about 600 megabytes of CD-ROM was required to produce the 32 volumes of the *Encyclopaedia Britannica* on a digital medium. This exact statement of data content tells us the limit of the CD-ROM. But there have been cases of people, gifted, but of about the same physical brain size as you or me, who could memorize pages of print that would dwarf the abilities of our finest computers. It is said, though there may be an apocryphal element to this, that Antonio Magliabechi (1633–1714) could routinely learn a book from a single reading.

Whether this is literally true is not so important as the consideration that he, like so many before and since, have never filled their brain to the point that no more information could be accepted. This is probably associated with the neural network, now so interesting to computer scientists. But the difference is that while computer neural networks are simple, our brains are not. On the principle that it is probably not possible to define either a theoretical limit to the human mind or its capacity to store information, using computer memory as an analogy for the brain is both superficial and misleading.

We should be clear, however, that the capacity of memory is not infinite. Within the finite volume of a head, the number of potential cellular interactions is finite, and so, therefore, is the amount of information we can hold. Neverthless, it is quite apparent that the amount most of us do hold is tiny when compared with our potential capacity, yet vast when compared with our computers. For example, we have a routine ability to process images in real time which takes the finest current computers many seconds to analyse. This is how a professional tennis player is able to calculate all the parameters of the flight of an approaching ball and move into the correct position to return it all within the space of less than half a second.

The digital analogy also runs into trouble when we consider environmental influences. Although DNA is the molecule that codes for all the proteins which end up as you or me, environmental influences also play an important part in development. As you grow and develop, so too do your potential and abilities, and this occurs apparently separate from the coding of your DNA. This is why our professional tennis player needs to train in order to realize his or her full potential.

We can conclude from all this that although DNA is so simple chemically it is capable of setting in train a remarkable series of events. This simple structure and highly conserved nature, through evolutionary time, makes it impossible to tell just by looking at a sequence of DNA from which organism it originated; not even whether it comes from a plant or animal. The only way to do that is to decode the DNA, bit by bit, and this is not practical in most cases, and even if it were, unless we know about the genes and their products we would still be ignorant of their origin. The fact that DNA varies very little from one organism to another is a reflection of the common nature of much of our biochemistry. If the enzymes we use are the same, it is reasonable to think that the genes coding them will be broadly the same as well.

Controlling gene activity

Something else which is apparent from studies of DNA is that as genes are switched on and off, so some way of controlling gene activity is essential. We cannot have random gene activity; for example, there would be no point in continuously producing a message for the production of lens protein in the eye. If that was to happen the eye would never be able to function correctly. Control is essential.

One method of control involves winding up DNA so that it is physically unavailable for use. This simple method of control can be seen in a basic form even in very rudimentary organisms like bacteria. Of course, when DNA is wound up it becomes compacted and potentially tangled. For this reason, instead of one continuous DNA molecule in organisms of great complexity, it is folded and refolded until a chromosome is formed which represents discrete and individually controlled units of inheritance. Luckily for the geneticist, at certain times during a cell's life cycle chromosomes are visible using a microscope and in some species, such as humans, considerable informative detail can be gained.

Although chromosomes are present in all organisms, the numbers present have no relationship to the complexity of the organism. Humans have 46 chromosomes, in 22 pairs with two sex-determining chromosomes, while the hermit crab is said to have 254 chromosomes. As one who has spent many years looking at chromosomes, I have to wonder as to the accuracy of this latter figure. Counting chromosomes is not easy!

Thus the DNA molecule is divided into discrete lengths, each being arranged into the structure called a chromosome. The structural importance of chromosomes is underlined by the very careful control which seems to be exercised over them by the cell. The whole of chromosomal control is a story of feedback and chemical self-regulation, with proteins coded for by the genes attaching themselves to the DNA and controlling the gene activity.

The most important of these control proteins are called histones. These are able to wrap DNA around themselves to form a sort of 'string of beads' appearance. As a starting point this is a most important step in forming a chromosome. The construction of a chromosome requires the string of beads to be wound up three more times, so becoming a very complicated coil of DNA and protein. This is best visualized by considering an elastic band wound between two points; as it becomes more and more wound up it progressively turns into a coiled coil and then a coiled coiled coil.

Chromosomes are superb pieces of packaging, taking single long molecules of DNA and winding them around proteins until the whole structure is visible as a single entity when viewed with a light microscope. At this stage the original DNA molecule has been packed in a ratio of 12,800 to 1. The importance of chromosomes as single units lies, primarily, in the need to control immensely long molecules of DNA during cell division. It would be a disaster for a cell if during cell division the material of heredity did not become distributed equally between the two products of cell division, the daughter cells, because it was all tangled up.

Just as the genetic code is universal, so, too, is the type of protein associated with chromosomes. Indeed, so important are the histone proteins that they represent some of the most highly conserved proteins between species and across evolutionary time. So highly conserved, in fact, that there is virtually no difference between histones found in humans and peas! Some genes code for products so important to life that they are almost impossible to change without affecting their activity. This is not a result of biology but something far more profound, reflecting the very basic uniformity in the chemistry of life – DNA.

THREE

Who are you?

DNA profiling and identification studies

If you wanted to know who somebody was and for some reason you couldn't ask him or her directly, what would you do? The answer is probably self-evident. You would simply ask someone who knows the person in question, assuming that you trust the person you asked to tell the truth. This apparently simple process of visual recognition, by you or anyone else, belies the complexity of what is actually going on when we recognize someone visually. Our facility to recognize people using visual clues, not just from facial features, but also from the way people stand, walk and move, is immensely sophisticated. So much so, in fact, that two people who have a close relationship can recognize each other at a distance when to anyone else they would just be a dot on the horizon. This process of recognition clearly uses clues so subtle that they defy quantification.

Under normal circumstances, we can recognize an approaching individual long before we can make out details of that person's face from their overall scale and the way they walk. But just how this is achieved and the neural complexities that go to make it work remains unknown. It is worth remembering that we, as animals, are not the best of animals when it comes to sight. That accolade probably belongs to birds of prey which need good eyesight not just to recognize a potential mate, but also to recognize a potential meal, which may be far in the distance.

To identify an individual categorically by sight requires knowledge of several different aspects, some of which are really quite nebulous and difficult to quantify, like the way a person moves or a skin tone. Take, for example, an attempt to describe hair colour; you may recognize the colour described, but our language is not adequate to cover all the different shades of hair colour that are found throughout the world. The main features included in our recognition routine are size, height, body language, overall appearance and, when appropriate, voice. All these things go together to identify a person. But what about identifying an individual from only a single character?

In many areas of modern life this question has attained exceptional importance, especially in areas related to security. In matters of state security, for example, it might be important to restrict access to certain information to a few, identified people. Perhaps access to valuable business information or equipment belonging to a commercial company needs to be restricted; this would require the categorical identification of personnel. It should be said that under these circumstances it is still not possible to invent a system better than the presence of a person who recognizes all the people with a right to be in a particular place at a particular time. Nevertheless, this solution is not always practical.

There are many different methods now being investigated to automate recognition for just these applications. These range from retina and iris scans to ear shape and palm prints. One of the problems with all these systems is that although these features are all unique to individuals, they can all be altered quite radically by disease or trauma. For example, the blood vessels of the retina can change with diabetes, or an ear or palm print can be changed by accidental damage. Besides which, if your hand is in plaster after an accident the palm would not be available anyway. Of course, all these techniques depend on having knowledge of each individual's features in the first place, so that a comparison can be made between them and a library entry known to be from that person.

In the case of crime, however, the situation is far more complicated. The likelihood of finding something at the scene of a crime that can be linked to a known individual is small. But sometimes material evidence is left at a crime scene, which can be compared later with a suspect who might have been arrested on other evidence. There are two types of evidence that do turn up at crime scenes quite often and which are of specific use and value to an investigating authority. The first of these is the well-known fingerprint and the second is a body fluid containing cellular material.

Fingerprinting

Fingerprinting is the oldest reliable method of personal identification known. Fingerprints have been recognized as unique for centuries and have been used in many different civilizations throughout history. One such use was the practice in China of placing a thumbmark on bank notes, half on the note itself and half on the counterfoil, giving a reliable method of matching the two.

The use of fingerprints for identification in modern times was first developed by a Scottish physician called Henry Faulds. While working in Tokyo during the 1870s Faulds amassed a large collection of fingerprints and because of his expertise in this field he was later consulted in the investigation of a crime where the thief had left a set of sooty fingerprints on a wall. When Faulds compared the crime scene prints with those of the man accused of the crime he was able to prove that it could not have been him. A later suspect, however, had fingerprints that matched the sooty ones. Following another successful identification at a second crime scene with a similar outcome, Faulds was convinced that fingerprints could be used for identification and he published his results in the journal *Nature* in 1880.

In order to make fingerprinting useful in criminal work, what was needed was a reliable system of classification. A remarkable scientist, Sir Francis Galton, produced such a system in the late nineteenth century. Born in Birmingham in 1822, Galton had the great advantage of independent wealth and, like many other eminent Victorians in this fortunate position, he was free to follow his studies and interests, which ranged over many different fields of science. Galton was particularly concerned that scientific data should be quantified as much as possible. At the time this was not always done, often simply because the techniques were not available to do it. Being a versatile thinker Galton was able to turn his mind to many different subjects in the search for methods of producing hard data. One of the first areas he worked on was weather maps, coining the term 'anticyclone'. In fact, Galton produced the first ever weather map, published in *The Times* in 1875.

Although less seen by the general public, probably the greatest scientific advance made by Galton was the correlation coefficient. This statistic is a numerical value ranging from 0 to 1 and is a measure of the relationship between two events. Zero is no relationship, while 1 is complete association. The meaning is rather more subtle than this, but it can be illustrated using a simple example. If you were to measure the energy consumption of a car against distance travelled on a flat track at a constant speed, it would be expected that energy consumption and distance would be closely correlated. In this case the correlation coefficient would be very nearly 1. Alternatively, a correlation between the driver's height and distance travelled would not be expected to show any relationship and would therefore have a correlation coefficient of 0.

To return to the story of fingerprinting, when Galton turned his mind to this subject he realized that if fingerprints were to be of any use, there needed to be some way in which the patterns and forms of the human fingerprint could be classified. What he eventually came up with was a system on which our methodology is based today. This was laid out in his book, *Finger Prints*, which was published in 1892.

It is an unchallenged idea that a fingerprint is unique; even the fingerprints of otherwise identical twins are different. Tiny variations in the fingerprint patterns, rather than ridge counts, can be used to distinguish between the prints of any two individuals. The inheritance of fingerprints is at least in part controlled by genes, under the influence of what is known as 'polygenic inheritance'. In polygenic inheritance several genes are involved in the control of a single characteristic.

We know that polygenic inheritance is significant in fingerprints from simple observations of what happens in chromosomally abnormal individuals. For example, in people with Down's syndrome, who have an additional chromosome 21, most fingers have a ridge pattern of loops rather than the more normal whorls or arches. Conversely, individuals with an extra chromosome 18, a condition sometimes called Edward's syndrome, have nearly all the patterns of the fingers in the form of arches.

It is clear that at the same time as the genetic control there must be some environmental factor involved as well, otherwise identical twins would have identical fingerprints. This environmental influence is probably small and occurs during growth and development in the uterus, but on something as subtle as a fingerprint a small variation can have a relatively large result.

It has been suggested that had fingerprint analysis been introduced a hundred years later, its acceptance would not have been so universal and complete. We now have a well-educated and critical general public who are much less likely to simply accept at face value what governments and their experts say. If invented now, analysis and interpretation of fingerprint results would be far more contentious, and quite rightly so. The difficulty of interpretation was underlined when, in 1995, the UK Home Office was forced to abandon an attempt to set up a national fingerprint database, simply because of technical inadequacies in quantifying what had always been seen as a definitive statement of individuality. One of the reasons that we can be sure there would have been more argument about fingerprinting if it had been introduced in the latter part of the twentieth century, rather than the nineteenth, is the strength of feeling voiced at the most recent introduction of a method to identify an individual for use in court: DNA profiling.

The alternative to fingerprinting – DNA profiling

Like fingerprinting, DNA profiling, or DNA typing as it is sometimes known, also uses a single character to identify a person, but in this case it is a far more complex character, the person's DNA sequence. In this technique cellular material containing DNA is used to generate a profile based upon measurement of small variations in specific areas of the genome. Originally this was referred to as a DNA fingerprint, but this name caused so much confusion in people's minds regarding its fallibility that it is now called DNA profiling. If we look at the history of DNA profiling we find that it did have a very auspicious start; the arguments started later.

In October 1986 a youth was charged with the sexual assault and murder of a teenage girl in Leicestershire, England. It was suspected that the same youth was also responsible for another sexual assault and murder that had been carried out some three years earlier. The assailant had left behind semen staining on the victim which allowed DNA to be used in a criminal investigation for the first time. Analysis clearly demonstrated two important points. The first was that the suspected youth could be completely ruled out as the assailant and the second was that, given the limits of the statistics, the two assaults were committed by the same person.

Consequently, in January 1987, the Leicestershire police force decided on an unprecedented action; they would ask all the males in the village concerned to voluntarily give blood samples so that they could be screened. The total number of people to be tested was approximately 5,000. Because of the size of the task, costs and time would be prohibitive if everyone was screened using DNA alone. It was decided, therefore, to make a preliminary screen of individuals using conventional blood grouping techniques. This is relatively quick and cheap, compared with DNA profiling, and would serve to eliminate the bulk of the male population from further investigations.

Preliminary screening left a core of about 500 samples which had to be tested using the much more highly discriminatory technique of DNA analysis. All the other samples had been ruled out using the blood tests. DNA profiling takes a great deal longer than simple ABO blood testing, as well as a great deal more manpower and facilities.

It was hoped that this final trawl through the remaining 500 samples would result in at least one person being implicated in the assaults and murders, but it did not. The person who eventually confessed to both crimes was found out because of his own bravado. He boasted in a public house of having paid a friend to give blood on his behalf. When he was eventually tested his profile matched those found at both of the crime scenes.

The use of DNA as a method of identifying an individual works in a fundamental and completely different way to all other methods of identification. Whereas earlier scene of crime methods relied on an image left behind, as in the case of a fingerprint, when DNA is used there is no remnant image, only cellular material; any image that might be used is embedded in the DNA. It may seem a circular argument, but since cellular material cannot be associated with an individual unless it can be compared with an individual, the presence of cellular material does not necessarily imply that the originator of it is the guilty party. In the case of rape, for example, it may be necessary to test the victim's partner to make sure that it is the assailant's semen that is being analysed.

In criminal cases, usable cellular material might be blood samples, semen or saliva, any part of the body, in fact, which contains cells with a nucleus, this being the main DNA-containing part of the cell. Not all cells contain a nucleus and therefore not all cells contain DNA; human red blood cells for example do not have a nucleus. Consequently, when a blood sample is analysed it is only the white blood cells, the lymphocytes, which are used for DNA profiling.

The identification of bodies

Almost as soon as highly variable sections of non-coding human DNA were discovered in the mid-1980s, the question was asked 'to what use can we put this DNA profiling?' Several answers emerged and although perhaps today the most familiar use is in the identification of criminals, one of the most useful early forensic applications was to help determine the identity of bodies. When a body is found with no immediate clue as to the identity of the person, traditional methods of identification involve comparison with reports of missing persons that fit the general outline of age and sex of the body. The person may have been the victim of an accident or murder, but of prime importance is to find out who they were.

Sometimes, however, traditional methods are just not feasible. For example, after an aeroplane crash from high altitude the bodies may be so badly damaged as to be barely recognizable as human and certainly impossible to identify individually by sight. Similarly, dismembered body parts are difficult to identify; if two legs and an arm are fished out of a river it is important to be able to tell whether they belong to one, two or even three people.

When a partially decayed body is discovered, rather than one unrecognizable due to physical trauma, the difficulties are the same but for different reasons. The main problem in identifying a long-dead body stems from breakdown and decay, which, unless the body is in a situation accessible to scavengers, is primarily caused by bacteria and fungi. Once decomposition has started it can be very difficult to do anything with traditional techniques of identification. The rate of decay in a wet and shallow grave, for example, can render most body material useless in a matter of years. In other situations the process can be even faster. In water, for example, skin very quickly discolours and comes off a body. So even detailed observations may not tell us very much about the individual, since skin colour and fingerprints will disappear.

For the same reasons traditional blood grouping cannot be done. The ABO blood groups are based on differences in certain proteins that occur on the surfaces of cells and these are subject to rapid decay in a dead body.

Fortunately, it takes much longer to destroy a molecule as robust as DNA, which is not only embedded deep in the cell, but is also very stable chemically. For this reason, DNA makes an ideal material for identification purposes. Although not completely immune to decay, DNA lasts extremely well, especially in cells which are protected by being deep within the body, like bone marrow. So well protected is this material that even in cases of exhumation after several years in the grave, DNA has been extracted and used for identification.

A slightly different set of circumstances is found in DNA extracted from ancient mummies. Just as food can be preserved by drying, so too can human bodies, especially when they are prepared with the care and expertise exhibited by the ancient Egyptians. In bodies prepared in this way, the preservation can be almost complete, with the cellular material untouched by decomposition and the DNA kept in remarkably good condition. Frequently, enough undamaged DNA can be extracted from a mummified body to enable the construction of a partial, if not complete, profile.

Under most circumstances, however, the breakdown of DNA in bodies can be a considerable problem in forensic science, especially in criminal cases where the disposal of the body, not preservation, has been the aim of the criminal. Decomposition takes place primarily as a result of the actions of bacteria and fungi. In fact there are very few conditions in which decomposition will take place without an external agent, such as bacteria or fungi. In the case of a body which has been eaten, for example, either cooked or raw, digestive enzymes will do exactly the same job as decomposition by microorganisms, the only difference is that it is an internal process.

Once the bacterial decomposition of a body has started, the DNA will start to break down and it becomes increasingly difficult to produce a proper DNA profile. The longer the DNA strand required for a profile, the more likely it is to have been damaged or destroyed by the action of microorganisms.

In 1981 a 15-year-old girl was murdered and her body buried, wrapped in a carpet. It was only in 1989, eight years later, that the remains were discovered and by that time they were, as can be imagined, in a dreadful state of decay. Although forensic scientists were able to reconstruct the facial features from the skull and dental records, the girl's identity was only confirmed by comparing remnants of DNA with that of other family members. This was only possible because the police had a good idea as to who she was and were therefore able to use DNA from members of her family for comparison. In cases such as these the body has to be aged and sexed, so that comparison with lists of missing persons can narrow down the search. Nobody could overestimate the distress all this causes to the families involved, but confirmation must be made, as much as anything, for the sake of families who are still searching for a missing loved one and who would do anything to know, one way or the other, where that person is or what happened to them.

By taking blood samples from other family members and building up a picture of the possible combinations of DNA bands that might occur in the person being investigated a picture emerges of whether they are related. If a match is found a probability can be generated which will indicate the likelihood of the result being found by chance, i.e. the likelihood that the person is not related but has the right profile by coincidence. As we shall see, there are many different methods that can be used to create a DNA profile and although the technologies are incompatible, they all produce a sequence of bands which can be compared between individuals.

DNA profiling can also be used not to determine a dead person's identity but to establish their connection with a crime after their death. In an unusual application of DNA profiling, the technology was used in one case to determine the guilt or innocence of a suspected murderer long after he had died and been buried. In fact, the suspect had died long before DNA analysis had even been devised. Exhumation was carried out on the body 16 years after burial so that DNA could be compared with material from an old crime scene. Although the results were sadly inconclusive in this case, it was possible to extract DNA from the bones, which was quite an achievement in itself. Unfortunately the DNA was just too badly damaged to be of use for profiling.

The development of DNA profiling

The scientific breakthrough that led to the possibility of DNA profiling as an identification technique was made by the British geneticist Alec Jeffreys in the mid-1980s. Jeffreys noticed that in any DNA sequence there were highly variable sections of DNA that did not appear to code for anything.

As is often the case in scientific research, Jeffreys did not set out to discover DNA profiling, this was a fortuitous observation made while he was searching for disease markers. A disease marker is a section of DNA which is closely associated with a gene, but not necessarily a part of it. Not in itself directly responsible for a disease, but being adjacent to the defective gene, a DNA marker is inherited with the disease. It is a distinctive and identifiable DNA sequence that can be determined and thus followed in a family. If the marker is transmitted to an individual, then so is the disease gene.

Of course, there will always be occasions when this will not work. Sometimes information is swapped between chromosomes in the natural process called crossing-over. If a swap is made between a marker and its associated defective disease gene then they will become separated. The further away two DNA sequences are on a chromosome, the more likely it is that crossing-over will take place between them and therefore render any attempt to link them together more difficult. When two segments or genes are on the same chromosome but widely spaced, they cannot be used as markers for each other, but when they are very close, and best of all, when they are next to each other, they form a linkage group in which the presence of the known sequence will always imply the presence of the unknown, disease, sequence.

This works even when the disease gene itself is unknown and where the marker is a non-coding region. A marker is often a single change in the DNA code which makes it susceptible to digestion by a specific bacterial enzyme. Most important in choosing, or discovering, markers is that they can be easily identified and are uniquely associated with the gene of interest. It would be of little use to identify a marker that was associated with the normal as well as the abnormal gene.

To return to the subject of DNA profiling, the potential of Jeffreys' highly variable non-coding DNA sequences was realized when it was discovered that although they did not appear to have any function themselves, they were unique to each individual. Except in the case of twins, each person has his or her own identifiable sequences. This provided a means of identification potentially more reliable and powerful than old-fashioned fingerprinting.

Methods of DNA analysis

There are three different methods used to analyse DNA for the purpose of identification. These are multilocus probes (MLP), single locus probes (SLP) and short tandem repeat analysis. The first two methods are broadly similar and are described here, while the third and most recently introduced is quite different and is described later on. All these techniques originated in biomedical research as methods of tracking disease genes in families, but they soon demonstrated their potential as tools in personal identification.

So what is it that these DNA profiling tests look at, and why are they so good at identifying an individual? The second question is easily answered: as more markers are used it becomes progressively less and less likely that two individuals will ever share the same combinations of DNA sequences. The first question, 'what is the nature of the variation looked at?' needs a little more explaining and takes us into the very depths of molecular biology.

The whole process of DNA profiling is dependent on the sites at which an enzyme, called an endonuclease, can cut a DNA strand.

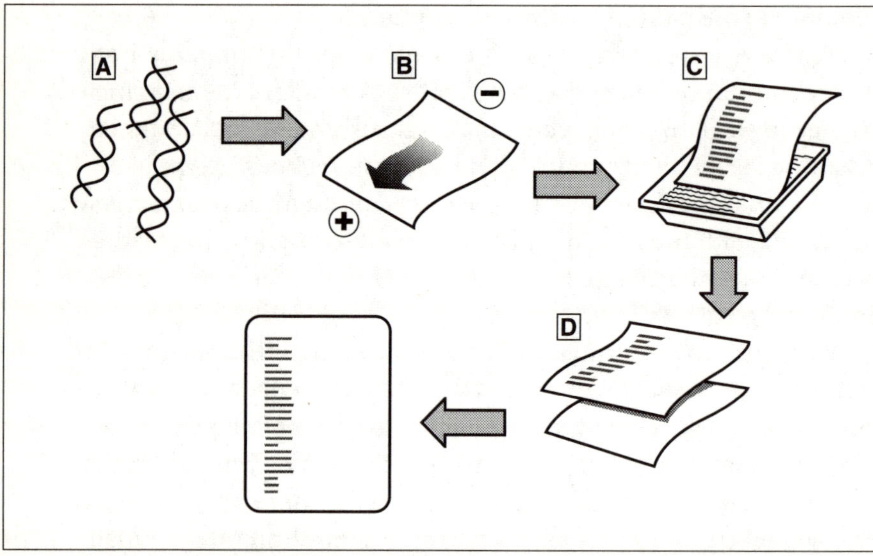

Constructing a DNA profile. (A) DNA is cut into short lengths using enzymes. (B) The different sizes of DNA fragment are separated on a gel under the influence of an electric current. (C) The gel is treated to stop the DNA passively diffusing within the gel. (D) DNA is transferred to a membrane where it can be probed with complementary sequences that have been tagged with either a fluorescent molecule or a radioactive label. This is then applied to a piece of X-ray film which gradually darkens in the areas in contact with the probe, resulting in a permanent record of the DNA profile.

Many enzymes of bacterial origin are capable of cutting DNA, and they always have a specific site for their action; usually a specific sequence of three to five DNA bases which they recognize. These bacterial enzymes are reputedly a natural defence against invading viruses. When a virus introduces its DNA to take over a bacterial cell, enzymes cut up the viral DNA making it inoperable.

Once a source of DNA has been located, the whole DNA profiling procedure, from extraction of the DNA through to completion, is carried out in a test tube. The process is outlined in the illustration on page 68. Incidentally, if you want to extract DNA yourself, a recipe can be found in the appendix to this book.

Once extracted, an enzyme is used to cut the DNA molecule into different lengths, but not randomly. As mentioned above, each enzyme cuts DNA at a specific site and the cuts are very precise, so that however many times the process is repeated, the DNA fragments from a particular individual will always be the same sizes. Next it is necessary to measure the sizes of the fragments. Because we are dealing with a collection of molecules it took considerable lateral thinking to work out a way to do this. It was finally achieved using a process called electrophoresis. When DNA is made to pass through a gel made of agarose jelly or a similar material called polyacrylamide using an electric current, strands of different lengths move through at different rates. Although the relationship between length and speed of travel is rather more complicated than a simple direct relationship, it is possible to make quite accurate measurements of DNA sizes from the overall picture which emerges. Part of the way in which this sizing is done is to run comparison pieces of DNA of known length on the same gel.

As would be expected, the smaller pieces of DNA travel faster than the longer ones. To get the DNA through the gel at a constant rate a steady electric current is applied to push the molecules through. Eventually, if the gel is long enough and the current is run for long enough, all the different sizes of DNA will have been separated into discrete bands. The longer the gel the greater the separation. Since the DNA fragments travel at their own speed, the longer a gel runs the greater the distance between the final bands.

At this point the DNA is invisible so it is imperative that it should be revealed; after all, if you cannot see the result there would be little point in starting the process. However, before the DNA can be revealed it has to be immobilized; if it is left on the gel it simply starts to diffuse in all directions. This is achieved by blotting the gel onto a nylon membrane, thus transferring the DNA onto a medium that will not allow it to move. This process is called Southern blotting, after its inventor, Edward Southern.

The Southern blot is now an exact copy of the gel, with the important distinction that once on the blot the DNA is fixed and immovable, whereas in the gel it is able to move freely, and if left would slowly diffuse and spread, making further analysis impossible. Because the DNA is now fixed it is quite easy to both handle and analyse the blot without the risk of the different pieces of DNA moving relative to each other, or worse still, coming off completely and ruining the result.

Once the DNA has been transferred to the blot a probe is prepared which has a DNA sequence complementary to a part of the DNA being tested. Because of the enormous potential for variation in the order of the four DNA bases, it is possible to generate a sequence many tens of bases in length, which will be quite specific as to where it will attach to the tested DNA. This section of DNA is referred to as a probe. If the probe is not specific, i.e. not a unique sequence, then it will attach to many different areas of the tested DNA. This is called a multilocus probe. If, on the other hand, the probe is so specific that it will only attach to a single area of the genome then it is called a single locus probe.

A multilocus probe will produce a final profile of many bands, while a single locus probe will produce a profile of either one or two bands. A single locus probe produces a single band if both copies of that sequence are identical. If two chromosomes in a pair, one inherited from the father and one from the mother, are slightly different, the result will be two bands in the single locus profile. If the two bands are the same size, then that individual is referred to as homozygous. If the two bands are different then the individual is heterozygous. Although multilocus and single locus probes are precise in the sequences to which they attach, these sections of DNA tend to be very large and are consequently rather variable in themselves. As a result, no matter what is claimed, the final analysis will always be open to subjective interpretation. This fact can, and has, resulted in some highly questionable results which have implicated individuals in crimes.

Once the DNA has been probed we still have the problem of visualizing the specific piece of DNA that we are interested in. There are several different ways of doing this. The earliest methods used a radioactive isotope, such as phosphorus, written as ^{32}P. Phosphorus is an important component of DNA, so substituting a radioactive version of the element does not alter its chemistry, but, of course, it does mean it has to be handled with extreme care. The DNA in which the radioactive phosphorus is incorporated can blacken photographic film, so once the radioactive probe has been attached to the DNA on the blot a piece of X-ray film can be laid over the top and left until the radiation has produced marks corresponding to the exact position of the probe.

The image produced using this technique is called an autoradiograph. It shows a series of tracks and bands exactly corresponding to those on the original gel, although it is two steps away from the original. The result of this long-winded process is the commonly seen banded pattern of a DNA profile. It must be emphasized that interpreting the banding pattern at all is difficult. While any interpretation is potentially fraught, the most contentious use of DNA profiling is quite clearly in a forensic context, not only in criminal investigations but also in resolving paternity issues. It is easy to see why this is an area brimming with argument. Not only is interpretation open to debate, but the implications of a mistake could be immense. Someone could be incarcerated for a very long time based on DNA evidence and if they are innocent it would not only be a gross injustice, but it would also imply that the true perpetrator is still free. Remember that DNA is not generally used in petty crime; its greatest use is in sorting out the results of the worst human behaviour imaginable.

Recently, partly due to the increasing awareness of the potential problems associated with radioactive material, methods of detection have altered radically. The method of choice now is chemical luminescence, which also has the advantage that it is much quicker than using radioactive probes. Chemiluminescence takes only hours, rather than days, to produce an image.

Chemiluminescence is created by attaching a molecule to the probe, rather than a radioactive atom. This makes the chemiluminescent probe larger than a simple radio-labelled DNA probe, but it does not alter the final result. The additional molecule is usually a specific enzyme which generates a signal, in the form of light, every time it comes into contact with a specific substrate molecule. The light is generated because as the enzyme splits the substrate, energy in the form of a photon of light is released. Breakdown of the substrate produces a very low intensity light, as can be imagined, so it would certainly be invisible to the naked eye at the rate at which it takes place. This is why a very sensitive photographic film has to be used, with an exposure of several hours.

The polymerase chain reaction

The most recently developed methods used in DNA analysis are so far removed from those original multilocus and single locus probes that the results are not in any way comparable. Although the new methods were originally introduced in order to increase both the precision and accuracy of DNA profiling, the two main issues of contention with the older methods, they also had a very important additional result: they enabled DNA profiling to be done using very small samples. It has even proved possible to produce a DNA profile from a single cell. Better even than that, when it was necessary to directly determine mutation rates in some genetic diseases it became possible to produce a profile from a single sperm, which carries only half the DNA of a normal cell.

This new process of analysis uses a technique called the polymerase chain reaction, normally shortened to PCR. PCR is a method of replicating DNA accurately in a test tube using naturally available enzymes (see illustration on page 73). Kary Mullis, the originator of PCR, first described the process at the American Society of Human Genetics in October 1985. Almost immediately the potential was recognized, not just for personal identification, but also for medical research and diagnostics. So important was this

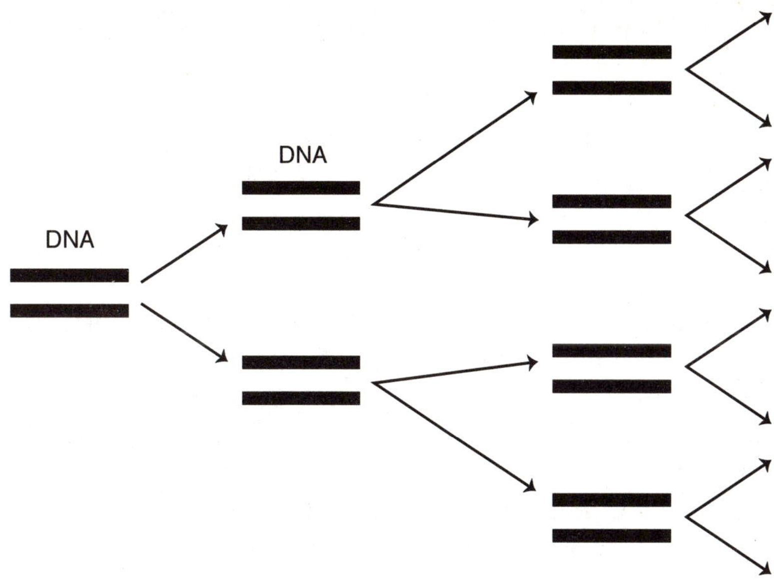

The polymerase chain reaction (PCR). In one of the most important recent developments in molecular biology, DNA is amplified using a heat-stable DNA polymerase enzyme and synthetic primers in the presence of a supply of nucleotides. Two primers, one for each strand, complementary to regions of the target DNA are added and the mixture is heated and cooled in a series of temperature cycles. Because the strands produced also act as templates for subsequent cycles, the amplification is exponential. It is therefore possible to start with a single piece of DNA and produce enough material that can be used for standard methods of DNA analysis.

new methodology that the award of a Nobel Prize to this brilliant, if idiosyncratic, scientist was given in 1993.

The replication of DNA using PCR relies on the extraordinary stability of certain bacterial enzymes; not the ones that cut DNA, this time, but the ones that replicate it. These enzymes, the DNA polymerases, are produced from bacteria living in very inhospitable environments, sometimes at temperatures and pressures that were once thought to be incompatible with life. At the bottom of the oceans there are deep fissures and vents which send water into the ocean depths at temperatures far in excess of 100^0C. It is only the phenomenal pressure that stops the water from boiling before it is diluted by the surrounding seawater.

These deep sea vents are the home for relatively simple creatures that seem to contradict the old idea that all life on earth, however far along the food chain it is, gets energy from the sun. Deep sea vent communities mostly contain species related to terrestrial forms, which would imply that they evolved from them. Energy seems to get into these ecosystems from microorganisms that utilize the chemical energy locked in the extremely hot seawater as it emerges from deep within the earth. The temperatures at which these bacteria thrive would kill a human, not only because of scalding but also because at these temperatures our enzymes would break down. Enzymes are proteins and when they are heated they coagulate and become permanently damaged. This 'denaturing' of proteins is the process that makes an egg go hard when it is cooked. It comes about because the three-dimensional structure so vital to the biological activity of proteins is changed by the action of heat.

Because of these extreme conditions, the bacteria living in deep sea vents need to have enzymes that are stable at very high temperatures. Scientists have taken advantage of this by making use of the high-temperature enzyme evolved by these bacteria to replicate DNA under laboratory conditions.

Although it is normal to think of ourselves as far removed from bacteria, clues to our common origins can be seen in genes and gene products. Looking at some gene products we may be surprised at just how similar they are across very different taxonomic groups. While some, ostensibly important aspects of life are surprisingly labile in genetic terms, others are not. For example, while the genes that code for digestive enzymes are highly variable, some genes are so well conserved as to be virtually indistinguishable from species to species. One such gene is that for a histone protein, the structural protein that is vital to the control of DNA and therefore the structure and function of chromosomes. Just as the basic structure of chromosomes is universal, as we discussed in Chapter 1, so too is the molecular make-up of the histone proteins. These proteins are so highly conserved through geological time that the difference between human histone protein H1 and pea histone protein H1 is minimal.

In the same way, enzymes associated with the replication of DNA are also highly conserved, and because the enzymes found in bacteria work just as well on human DNA and plant DNA, this gives us a method of artificially replicating DNA in the laboratory. By using DNA polymerase from a bacterium which normally lives in the scalding hot waters of either hot springs or hot deep sea vents, we can be sure that the enzyme will remain stable through many cycles of replication without denaturing due to heat. The only other thing required is a primer complementary to part of the DNA strand.

As a result of the discovery and development of PCR using these bacterial enzymes, it became possible to replicate DNA not only from very small samples, but also from DNA which had started to break up. The process of PCR requires a bacterial enzyme, DNA polymerase, and two primers. A primer is a sequence of DNA which, like a probe, is complementary to part of the sequence in which we are interested. Primers are usually about 20 bases long, long enough to be specific for that sequence of DNA in that organism. One primer sits at one end of the target DNA and the other primer at the other end. The gap in the middle is then repeatedly replicated by the enzyme. By heating the DNA it can be made to unzip down the middle, so that when it cools, instead of reforming with its complementary strand, the primer attaches instead. At this stage the DNA polymerase enzyme gets to work, adding complementary bases to the strand, until it is reheated and the new strand unzips, so giving access to another primer for replication. This process is, quite literally, carried out in a machine whose sole purpose is to heat up and cool down the test tubes.

The net result is an exponential growth in the amount of DNA. If you were to start with a single molecule of DNA then after one round of replication you would have two molecules, after the next round you would have four, then eight, sixteen, 32, 64, 128, 256 and so on. From a single piece of DNA it is possible, with careful control of conditions, to replicate enough DNA for a complete analysis.

Following the replication stage, once again it is necessary to run a gel to separate the various bits of DNA so that detection of the products will give us a measure of the size of the fragments. It is the fragment size that is crucial here. Originally, sizing of the fragments was done in much the same way as with single locus and multilocus profiles, as described above. But now new technology has been developed to take advantage of the very high precision of PCR reactions. This takes the form of a fluorescent tag attached to the primer so that when illuminated by a tuned laser it emits a coloured light. This is all done automatically and the data passed directly to a computer. There is no longer a piece of film as a record of the gel, just an image in a computer.

Short tandem repeats

With advances in modern technology, the way in which DNA is analysed has totally changed in the last decade. No longer is the DNA cut and measured approximately, it is now measured in units of repeats, with very high precision, down to the last base. These units of repeated segments of DNA are called short tandem repeats, or STRs. As we know, DNA is constructed from only four different chemical units, the four bases described in Chapter 1. An STR is a sequence of DNA which can be represented by those four bases. For example:

CGAC CGAC CGAC CGAC CGAC
GGGA GGGA GGGA GGGA GGGA
CGAT CGAT CGAT CGAT CGAT

are all short tandem repeats, each letter representing one DNA unit.

STRs vary in size – some contain four repeats or less; some may contain six, or even more:

CGAT CGAT CGAT CGAT CGAT CGAT
CCAG CCAG CCAG
AATC AATC AATC AATC AATC AATC AATC AATC AATC.

The repeating motif of a specific STR is always the same, but different STRs have different motifs.

The length of these short tandem repeats varies between individuals. So, by measuring the number of repeats in a sample, and comparing it to a suspect, a match will either be found or ruled out. Similarly, if bits of a body need to be identified as coming from the same corpse STR analysis can be used, even when the DNA is starting to break up.

In order to confirm a match, several different STR systems need to be combined to produce a composite picture built up of different-sized repeat sequences. This is because although there is variation in the number of repeats, the variation is relatively small, so a large number of STRs need to be looked at to give a clear answer to the question 'is this the person we are looking for?' With each new STR that is measured, the probability of a chance match becomes smaller.

The picture for the scientist is further clarified by another characteristic of short tandem repeats: for every STR an individual will have two versions, or alleles. Each allele is inherited from a parent, one from the mother and one from the father. So it is also possible to use this technique for analysis of parentage. In looking at a child's STR profile, those bands that do not come from one parent must come from the other. An individual's alleles may be the same, in which case they are called homozygous, or they may be different, in which case they are heterozygous.

DNA profiling and the law

The question remains, of course, whatever method is used to analyse DNA, what is done with the information generated and what can we learn from it? This clearly depends on the context: in the case of a crime, the procedure is slightly different to that used in a disputed paternity. But one thing can be said with complete certainty – it is possible for a DNA result to eliminate a person either as a parent or as a criminal without a shadow of doubt, but it cannot categorically demonstrate that someone is a criminal or the disputed parent. Finding a match is termed an 'inclusion', but an inclusion can only give an indication of the probability that two samples originated from the same individual. In a criminal case an inclusion most definitely does not give a probability of guilt or innocence.

When DNA is analysed for use in criminal cases it is normal to have a sample from the scene of the crime, with which a sample from a suspect can be compared. It must be understood that although it is sometimes claimed as enough to find a match, this is far from the end of the story. There have been, admittedly rare, random matches found which demonstrate the fallibility of DNA profiling. Consequently, every method of producing a DNA profile, however it is done, has to have an associated database of test samples analysed in the same way, which gives the frequency of the profile in the general population. It is this figure which results in misunderstandings about DNA evidence.

In the early days of the use of DNA profiling in legal cases, an American court was given a figure of one in 738 trillion for the probability of finding a specific profile by chance, rather more than the population of the planet. This figure is the frequency with which such a profile would be expected to appear in the general population. It most definitely is not the probability of innocence or guilt. Undoubtedly the biggest problem with these figures is just how difficult they are to comprehend. Beyond a certain point, depending on culture and background, numerical values just become perceived as 'a lot'. As an attempt to help in understanding these figures the UK Home Office now uses a method of interpreting the numbers, using a scale such as 'weak', 'good', or 'very good', and if pressed for a figure relating the frequency of the profile in the population will round them down so that they are of a more reasonable and understandable magnitude.

Precisely what weight is given to such statements depends on ancillary evidence presented to the courts, and it should be remembered that it is for the court to decide what importance to attribute to evidence, it is not the job of the prosecution or defence. It is also, of course, for the court alone to judge guilt or innocence, whether they get it right or wrong.

In paternity disputes the situation is rather different. In the first place, such cases are nearly always paternity disputes, rather than maternity or parentage disputes. This is because, although it is possible that such a case could arise, perhaps when a mix-up happens in a maternity department, generally speaking it is known who the mother of a child is. On the paternal side, witnesses to the conception are rare and multiple partners not as uncommon as one might imagine.

When paternity is disputed, a chain of events can be set in motion in which using DNA to help clarify the situation may be pivotal, but not always necessary. In the simplest situation a blood sample could be looked at for the ABO blood groups. Although not particularly discriminating, ABO blood groups are controlled directly by genes, so although it is the protein product of the gene which is assayed, it is only one step away from the genes. Like all such methods it is possible, using ABO blood groups, to categorically prove non-paternity, but not to categorically prove paternity. When direct analysis of DNA is brought into the equation, the discriminatory power is greatly enhanced and probabilities of paternity can be stated far better, but still not with absolute certainty. This residual doubt is associated with the possibility that another person may share the same blood group or DNA profile.

The manner in which DNA is used in paternity is essentially the same as in criminal cases, except that more samples are needed, these being from the mother, child and suspected father, or if there are two males in dispute, from both of them. If the DNA profiles of the child and the mother are known, whether by using single locus or multilocus probes, or STR technology, it should be possible to establish the probability of an individual being the father of a child. This is the technique used in the UK by the Child Support Agency to ascertain paternity. Whichever method is used, the one thing that is certain is that if the alleles of the child cannot be accounted for by a mixture of the alleles from the mother and the father, then, assuming we are certain who the mother is, the father can be ruled out categorically. If, on the other hand, an individual cannot be ruled out, it is, once again, necessary to go back to statistics. After all, if the DNA profiles of the child and the parents were as common as, say, brown eyes, then trying to sort out a disputed paternity based on a DNA profile would be worthless.

Do not forget that a match is not in itself adequate to determine a relationship, whether between a scene of crime specimen and a suspect, or in the interpretation of a sample in a disputed paternity between a father and child.

The last of the Romanovs

Where questions of paternity relate to the inheritance of both title and wealth, it is sometimes necessary to take unusual steps to elucidate what would otherwise be an intractable and unresolvable problem. It is here that some of the very robust DNA from mitochondria as opposed to nuclear DNA can be used, especially if an exhumation is needed. A spectacular case that made use of mitochondrial DNA to answer such a question concerned the identification of the last of the Romanov family, part of the royal lineage of Russia.

In 1990 the claim of Anna Anderson to be Anastasia, youngest daughter of Tsar Nicholas II, which had been widely debated since the disappearance of the Russian royal family during the revolution of 1918, suddenly hit the headlines again. After a great deal of searching old records held in Russia, a research team had discovered a shallow grave in the eastern foothills of the Urals near a town called Ekaterinberg. The grave contained the badly decomposed bodies of nine individuals, and even in this state it was obvious that these were victims of murder. Given that it was known that the Romanovs had died by firing squad on 16 July 1918 at Ekaterinberg, it seemed probable that these were indeed the bodies of the Tsar and his family.

According to local records, the bodies had been unceremoniously dumped down a mineshaft in the first instance, but then two days later were moved to the shallow grave. It was perhaps inevitable that the discovery of this grave decades later would receive worldwide attention and that the most sophisticated techniques available to contemporary pathologists would be brought to bear on the subject. Not only was it of great public interest to ascertain whether these were indeed the bodies of the Russian royal family but it was also of political and financial importance to substantiate or disprove the claim of Anna Anderson to be a member of that family.

It was decided to conduct DNA testing on the bodies after the discovery of the grave but, as one would expect, the more usual techniques of DNA profiling were completely useless because the nuclear DNA was badly damaged after 70 years in the ground. The only alternative was to use a slightly different source of material called mitochondrial DNA.

As we have mentioned earlier, not all the DNA in a cell resides in the nucleus; a second type of DNA exists in the subcellular bodies called mitochondria. Mitochondria are the site of energy production in the cell and are thus very important in metabolism. It is thought that they originated as remnants of symbiotic bacteria which invaded cells in the dim and distant past of prehistory, before the arrival of multicellular organisms. The unique property of mitochondrial DNA is that in humans it is generally thought only to be inherited from the mother's side of the family. This gives a perfect method of tracing descent through a maternal line.

Mitochondrial DNA is found in the form of a sort of distorted circle of about 16,570 bases. It is thus much smaller in quantity than the nuclear DNA but it also differs in another way that is particularly pertinent to its use in identification. Strangely, mitochondrial DNA has areas which can vary *within* an individual, not just between individuals. This clearly impacts on its use in DNA profiling; it is sometimes not only insufficient to say that a match between individuals is significant, but it might be positively misleading.

As a subject for analysis, mitochondrial DNA has the advantage that it survives the ravages of decomposition better than nuclear DNA. This is not because it is more robust than nuclear DNA, but simply that the mitochondria give it better protection. The cells where mitochondria are most numerous, within bone, for example, also tend to be well protected. It takes much longer for bacteria and fungi to penetrate into the centre of bones than soft internal organs and muscles, so mitochondrial DNA can last very much longer in these well-protected internal parts than elsewhere. Mitochondria are found in large numbers in all cells, with the exception of red blood cells. Mitochondrial DNA is as easily damaged as nuclear DNA in soft tissues, like skin and muscle, but when it is lodged in bone it can remain protected for many years.

To return to the grave of the Romanovs, after much investigation it was decided that it did indeed contain the bodies of the last Tsar and Tsarina of Russia, three of their daughters and four other adults. The three daughters were identified as Marie, Tatiana and Olga. By deduction, two members of the family were absent – the son Alexei and Grand Duchess Anastasia.

Although questions regarding the identity of the unknown adults in the grave were asked, most people were more concerned with the question of what happened to the bodies of the missing children. Alexei was known to be a haemophiliac and nobody came forward claiming to be him, so it was assumed that he, too, had died. Interestingly, due to the well-kept records of the European royal families we know that his haemophilia was inherited matri-lineally from his great grandmother Queen Victoria of England, within whom the original mutation took place. We return to this theme in more detail in Chapter 5 on DNA and disease.

So what had happened to the young Grand Duchess Anastasia and what of Anna Anderson's claim that she was Anastasia? This claim was originally made by Anna when she was in a Berlin mental hospital in 1921, but there was no means of proving it one way or the other during her lifetime. Even at the time of her death in 1984 in Charlottesville, Virginia, in the USA, there was no reliable analysis of DNA that could be used in the case. However, in 1979, Anna Anderson had been taken into the Martha Jefferson Hospital for surgery where a small bowel biopsy had been taken. This had been embedded in wax and preserved – a standard procedure for a tissue biopsy destined for use in microscopy. Thanks to the robust nature of DNA this sample was ideal for analysis many years later, yet even though it was badly degraded by time and by the method used to embed the specimen in wax.

Using the short tandem repeat analysis described earlier the debate over Anna's claim was silenced once and for all. It was shown that surviving DNA from the bones in the grave was inconsistent with Anna Anderson being a member of the Romanov family. Moreover, a comparison between mitochondrial DNA from Anna Anderson and that from the Duke of Edinburgh, a known direct maternal descendant of the Tsarina, was also conclusive. Had Anna and the Duke been maternally related, the comparison would have given identical results. The fact that it did not was conclusive proof that Anna was not the Tsarina's daughter.

In the event, Anna Anderson's true identity was also revealed by the use of mitochondrial DNA analysis. It was suggested that she was, in fact, a Pole called Franzisca Schanzkowska. A documented great-nephew, maternally related, of Franzisca Schanzkowska was located and his mitochondrial DNA matched that of Anna Anderson. Although this latter evidence does not, in itself, constitute proof of her real identity, we can say categorically that DNA demonstrated she was not a Romanov.

Thomas Jefferson

More recently, DNA analysis has been used to shed light on an even older puzzle, involving the third president of the United States, Thomas Jefferson. There was a suggestion made in 1802 that Thomas Jefferson had fathered at least one child with a slave called Sally Hemings. Clearly we will never know the exact nature of the relationship between these two people, but some evidence as to the outcome of their supposed sexual encounter has survived. In this case it was by looking at Y chromosome variation that some part of the story was revealed.

Unlike mitochondrial DNA, which is only inherited via the maternal line, Y chromosomes are only inherited from the father. They are also quite robust in descent, not normally being involved in swapping genetic material with any other chromosome. As a result, Y chromosome DNA tends to pass down through generations unaltered. This means that by looking at variant DNA sequences on the Y chromosome we can learn something of the lineage of a male line of descent. In this particular case Y chromosome DNA known to be descended from Field Jefferson, a paternal uncle of Thomas who would therefore share the same Y chromosome DNA, could be compared with male descendants of Sally Hemings. What was discovered was that, unless there was some illegitimacy further down the line, Thomas Jefferson was the biological father of Eston Hemings Jefferson, the mother being Sally Hemings.

However, plantations at that time were inbred communities and mixing of different racial groups seems to have been common. Indeed Sally Hemings was a light mulatto, thought herself to be the half sister of the wife of Thomas Jefferson. It was also common for slaves to be passed on just the same as any other property. So, any other male descendant would have the same Y chromosome as Thomas Jefferson and this includes his brother Randolph and his five sons. So although the question must remain open, in taking the balance of probabilities it would seem likely that Thomas Jefferson was the father of Eston Hemings, but it is neither proved nor disproved.

Although when using DNA to test questions of personal identity a negative result has far more value than a positive result, such an outcome can seem unsatisfactory. But consider this; ruling someone out of a criminal investigation or a paternity dispute can in its own way be positive. DNA has undoubtedly become the most powerful tool in the armoury of methods with which to tackle the difficult problem of identifying individuals. But it is a mistake to assume that a simple match means a proven identity. Any profile can be assigned a frequency in a population. If a profile is so common as to appear in the majority of a population then, just like standard ABO blood groups, a match is of little use in clarifying a complicated situation.

FOUR

From dinosaurs to quaggas

The role of DNA in evolution

When scientists first recognized that the relations between species as described by Victorian naturalists in the nineteenth century were, generally speaking, reflected in their genes, it came as quite a surprise. Although it is not always the case, some groups of animals show remarkable physical similarities, and can be instantly recognized as belonging together. One such group is the parrots. These range in size from the six tiny species found around New Guinea and the Solomon Islands, some of which are less than 10cm long, through the familiar budgerigars of Australia to the enormous flightless kakapo of New Zealand and the macaws of South America, which are nearly a metre in length.

Whatever their size, the thing all these birds have in common is that they all look like parrots. If DNA suggested that, say, African grey parrots were not true parrots, we would be extremely surprised. We might even think that the DNA analysis is flawed and this would be quite reasonable. But sometimes the relationship between different species is difficult to see. Under these circumstances DNA may well be able to clarify relationships between taxonomic groups. We might even think that the DNA analysis is flawed, and this would be quite a reasonable conclusion to arrive at. But sometimes, certainly to the untrained eye, the relationship between different species is difficult to see. This is the case with tomatoes, peppers, aubergines, potatoes and deadly nightshade, all of which are in the same family of plants.

Phylogeny and taxonomy

Without a doubt, knowledge of DNA has elucidated some extraordinary relationships among forms of life on this planet. Not least of all taxonomic and phylogenetic relationships. Phylogeny is often seen as the same as taxonomy, but it is not. Phylogeny describes the relationships between groups reflected by evolutionary history, whereas taxonomy is the study of the classification of organisms based, usually, on physical appearance. Put another way, phylogeny represents the relationships between ancient and new groups of organisms such as reptiles and birds, while taxonomy relates species to each other.

The relationships observed by early taxonomists have been reiterated again and again in the DNA record as viewed by molecular biologists. But do not be fooled by this. What we see in the genetic record now was perfectly apparent to those taxonomists who worked so many years ago. The difference between then and now is really only one of scale; instead of looking at the molecular relationships they would have looked at gross morphology.

These earliest taxonomists were, indeed, looking at genes, but at a level far removed from direct analysis of DNA. Needless to say they did not realize that the characters they were looking at to construct their taxonomies were inherited and the direct result of DNA and genes. It is important to remember the context within which the first taxomonists were working. At this time the teachings of the Christian Church were quite clear: species were created whole and were immutable, and there was certainly no question of species evolving new forms. Given this background, biologists naturally assumed that what they were looking at were static and unchanging species. Therefore the relationships being described were not seen as phylogenetic, but simply as descriptions of relationships between extant species: a taxonomy.

Because the analysis of proteins predated direct analysis of genes, this was the first area of molecular analysis that was applied to relationships between species. Thus the phylogenies and taxonomies first put forward by morphologists were looked at later by biochemists investigating proteins. In general, the same relationships were discovered. More recently, when known genes and stretches of DNA had been sequenced and were available for this sort of analysis by comparison between species, they quickly became the subject of taxonomic investigation. Even though, very often, no protein product was known from these stretches of DNA and it was only possible to compare sequences, it soon became apparent that these analyses paralleled protein structures, and the link between DNA and the phylogenies produced by those careful taxonomists of previous centuries emerged.

This sequence of events carries a very important point which should not be dismissed lightly. It is one of the best and most readily seen direct indications that genes reflect proteins and proteins reflect morphology. Because these three systems of investigating relationships can be, and in fact usually are, carried out independently, the results reinforce each other and the emerging picture of evolution. We can be fairly sure, then, that genes are at least in part responsible for morphology and that they reflect the relationships between different groups of organisms.

How many species are there?

When asking just how closely two species are related, this question has to be looked at either in the context of the length of time that has elapsed since divergence or with respect to the other members of the same group. For these investigations molecular biology has the advantage of shifting the skill away from traditional taxonomy, where interpretation is difficult, contentious and subjective, towards a more objective interpretation. This is not to say that taxonomists are redundant, but their skills have been augmented by DNA analysis at the finer end of the scale of phylogenies – the species. Defining a species is hard enough, but with the continual interest in biodiversity it has become important to be able to at least attempt to say how many species there are extant on the earth.

Many estimates have been made, but one important fact needs to be borne in mind regarding the expression of diversity: there is not, nor could there ever be an infinite variety of species. Infinite means without end, which is quite different to 'a very large number'. And it would also be wrong to use the number of identified species as a guide to the total number of species because those groups which are most fashionable will have the largest number of known species. When the great biologist J.B.S. Haldane was asked what a lifetime in biology had taught him about God, he is said to have replied 'He has an inordinate fondness for beetles.' This is a reflection of the huge numbers of described species of beetle that there are. It is a commonly known fact that the insects currently outnumber all other species put together, but it is also true that the beetles are the largest group of insects. Nevertheless, there is no particular reason why beetles should outnumber all the other insects, it is most likely a simple reflection of the activities of the early taxonomists.

Some groups of organisms are sufficiently well known to be almost completely counted, with the proviso that we can never be sure that all species have been discovered. For example, there are 9,881 bird species and 4,327 mammal species known, out of about 1,770,000 described species and a total estimated number of between 3 and 5 million.

When we want to know about relationships between different species the first thing we do is look at shape, the physical structure of the organism. Size and colour tend not to be so informative as shape and structure. For example, by counting the number of legs on a creature it is easy to tell whether it is an insect, which has six legs, or a spider, which has eight. If one looks at colour, on the other hand, there are innumerable species that share the same colour. If we used colour to define groupings, we might conclude, for example, that green beetles, green fish and green birds were all closely related to plants.

The case of the marsupials

So it is shape and structure that tell us most about species. We can use this observation to do many things from the simple job of recognizing a bird as a bird, to the more complex task of identifying a skull as that of a marsupial. This latter example provides us with a fascinating and instructive case.

There are a series of identifying characters which can be used to differentiate a marsupial from a placental mammal. Most people think of marsupials as being generally similar to placental mammals but different in that they carry their young in pouches. But this is not the whole story; differences between placental mammals, like you and me, and marsupials are reflected in all manner of different features, including distinct characters identifiable in the skeleton. These differences extend far beyond simple changes in dentition or diet. Indeed, there used to be a thriving group of marsupial carnivores, now sadly nearly all extinct, as well as grazers like kangaroos and wallabies. But the important thing is that they were all recognizable as marsupials, reflecting the evolution of this group having being independent of placental mammals for many millions of years.

After the early mesozoic period, which started about 230 million years ago, two major land masses were formed. These were Gondwanaland in the south and Laurasia in the north, separated by the Tethys Sea. This sea presented an impassable barrier to land animals migrating between Gondwanaland and Laurasia. It is most likely that marsupials originated in what is now South America, spreading northwards into North America and Europe and southwards via Antarctica to Australia. Competition with placental mammals is generally seen as the reason that marsupials are now exclusively confined to South America, where there are 77 species, and Australia where there are about 161 species remaining. In Australia, marsupials and egg-laying monotremes, such as the duck-billed platypus, were the only mammals present until humans arrived, first introducing dingoes and then all manner of other non-native species. This means that the marsupials of Australia, unlike those of South America, did not have to compete with placental mammals before the arrival of humans.

Some of the unique features of marsupials are evident in their skeletons and there are many pointers that the experienced observer can use to decide whether a skull belongs to a marsupial or not. For example, marsupials generally only have one complete set of teeth and only one tooth is replaced. The bony palate in the roof of the mouth is also incomplete, although this is seen in other 'primitive' mammalian groups as well, such as insectivores.

When it comes to the rather more objective analysis of DNA from marsupials, sometimes it helps and sometimes it does not. Taken across the broad spectrum of marsupial species there is a wide range of DNA contents. The amount of DNA found in an individual cell of any species varies from less than that found in humans to nearly one and a half times the human DNA content. As we pointed out earlier, the amount of DNA found in the nucleus does not reflect an organism's complexity. For example, some salamanders have up to 20 times the amount of DNA found in human cells. The additional material is simply redundant, though how and why it got there is unknown. Interestingly, it is not so much the DNA content that can help us to elucidate evolution amongst marsupials, as the arrangement of DNA into chromosomes. For example, there are several distinct species of rock wallabies from Australia that, although closely related, have widely differing numbers of chromosomes. It has proved possible to create a lineage of how this occurred, starting from the original premise that, broadly speaking, rock wallabies have the same DNA content no matter what their species is. The differences in DNA between species are not simply ones of DNA sequence, but of the manner in which the DNA is arranged into chromosomes. With rock wallabies, it has been possible to take the chromosomes, stain them so that they show their internal structure, and then ascertain which chromosomal sections are related.

It would seem that the basic chromosome number in this group is 22, but many rearrangements have taken place resulting in reduced chromosome numbers. This has taken place by chromosomes becoming physically attached to each other by their ends. By going back to what is thought to have been a primitive chromosomal constitution for rock wallabies and taking tissue samples, it has proved possible to demonstrate an interesting phenomena in this group of animals. When tissue samples are grown in vitro in the laboratory it is possible for them to undergo changes which would not necessarily be compatible with life in the complete organism.

In the case of rock wallabies it is clear that the separation into different species has proceeded by modification of chromosomes, at least as much as by changes to the DNA within them. What became apparent from tissue culture was that the areas of chromosomes given to breaks and rearrangements *in vitro* are the exact same sites where rearrangements have taken place amongst rock wallabies in the wild, many millions of years ago. It seems reasonable, then, that although evolution requires changes in DNA on quite a large scale, the formation of separate species can take place by very small changes in DNA, but quite large changes in its organization in the chromosomes.

Changes in taxonomy

Traditional taxonomies and phylogenies produced by different workers have always been remarkably consistent. It has to be said, though, that most biologists really do not have the skills to accurately describe and interpret an independent taxonomic relationship between diverse groups. The system we now have was developed over several hundred years, as more information regarding interrelationships was added to the overall picture. The extreme example of this understanding must be that of Charles Darwin, who synthesized from these simple taxonomic relationships something far greater: the idea that one species could generate another.

The traditional manner in which changes in taxonomic groupings were achieved was, without doubt, slow and laborious. It would involve a biologist with a particular interest in a group of plants or animals developing an understanding of how the taxonomy that was currently used had been changed and developed as more information was accumulated. Of course this also included the discovery of new species, for these were often the intermediates linking one family or group of species with another. Should a noticeable discrepancy appear in the mind of the scientist regarding the group being worked on, then perhaps a step would be made to redefine the relationship between organisms. This would then become the acceptable taxonomy, relating different groups together.

A change to traditional taxonomies is never undertaken lightly. Reputations have been made and lost by precipitous publication of reworked taxonomies. The refinements which occur from time to time in scientific literature rarely have the effect of redefining large groups, tending to be small changes which gradually build into a clearer picture of the relationship between species. In today's scientific climate it is probably even less likely that any revolutionary ideas in taxonomy will be produced and accepted. Taxonomy is no longer a fashionable area of research and there is a mistaken belief that taxonomists are not necessary for modern biology.

Probably the greatest taxonomic revolution in recent times was the change in thinking that resulted in the recognition that avians, birds, originated from a reptilian stock. Rearrangements of such magnitude in phylogenetic relationships are rare, and in the case of birds and reptiles the original idea was based on morphology and implied physiology, such as bone and blood vessel structure. It was certainly not worked out using DNA.

Using DNA to define and identify species

What is possible with the analysis of DNA is that subtle changes can be made to our understanding of the relationships between species. In this respect DNA helps taxonomists working within already well-defined groups. This may not in itself appear significant, especially to those who think taxonomy is nothing but a sterile activity involved in shuffling indefinable species around a librarian's card index. But nothing could be further from the truth. Taxonomy and phylogenetic relationships are pivotal to the understanding of life around us and identifying species is essential before we can even start this process.

Take, for example, the case of the mosquitoes that carry the disease malaria. These are of the genus *Anopheles*, but if you want to develop either better anti-malarial drugs, or better methods of control of malaria-carrying mosquitoes, you really must know how to identify your foe more specifically than this. Very few mosquito species are able to transmit malaria and lack of knowledge has resulted in many abortive research projects ending prematurely because of a fundamental inability to identify exactly which species was being investigated.

In the genus *Anopheles* there are at least twelve species which can spread the disease, of which four species carry four different types of malaria. If you take a closely related and morphologically very similar mosquito genus, *Aedes*, the story is the same, only here we have seventeen species, and these spread a different set of diseases from yellow fever to Murray Valley encephalitis and Dengue fever. It takes a practised eye to determine which species you are dealing with and if you get it wrong a great deal of time and energy will have been wasted trying to develop control techniques for the wrong disease vector.

An extreme case of species ignorance I know of involved mistaking a human for a mouse! A research project, based on a specific mouse cell line, was ruined because it was eventually discovered that the cells being used were not mouse cells but human cells. This apparently ridiculous confusion took place because cells were being used rather than whole organisms. But surprising as it may seem, this is quite an easy mistake to make: once a cell line is in culture one bunch of cells looks much like another. It is only by looking in detail at cell surface proteins, chromosomes, or sequences of DNA that it becomes possible to identify with any certainty which species the cells originated from.

In conclusion, we can say that the one thing we need to know for sure is the correct identity of any organism that we are dealing with. In this respect taxonomy becomes very much a case of agreeing between us exactly which animal or plant we are talking about. After all, if you and I do not agree exactly what we mean, for example, when we say 'toadstool' there is no way of knowing which particular fungus we are dealing with – it may be edible and it may be poisonous. It is for this reason that Latin names are used in science to identify plants and animals. These have not been dreamt up to make life difficult for us by being mysterious or arcane, they are used because they can be understood internationally and without confusion.

Refining the techniques

When looking at smaller differences between species, and as far as we humans are concerned these may be just as important as large ones, we may need more accurate tools than those available to Victorian naturalists. Morphology might just not be enough on its own. It is surprising that so much detailed and accurate work was carried out using morphological features. In plants this would usually be from flower structures, while insect genitalia and feeding behaviour provided entomologists with subtle differences between species that would otherwise go unnoticed. One area where the use of morphology alone becomes more difficult is when looking at relationships between distinct and distinctive groups. An example from the plant kingdom illustrates this problem.

Monocotyledons and dicotyledons are the two major groupings of flowering plants, the name referring to the way in which they start growing from seed. Now, in the monocotyledons there is a group of plants which is regarded as one of the most highly adapted of all plant groups and one of the most easily recognized – the orchids. Most orchids have flowers that are immediately obvious, so that even if we cannot recognize an orchid from its foliage, we can always recognize its flower. So, since an orchid is an orchid and does not resemble anything else, the questions that arise are what is their origin and to which plant group are they most closely allied?

Questions such as these are asked about all manner of other plant and animal relationships between groups that diverged many millions of years ago, and morphological data do not always help in providing answers. Groups that diverged long ago in geological time are seen now in 'snapshot' form, that is with no extant intermediate species, so determining the relationship between two different groups can be extremely difficult. Into this arena, molecular biology and DNA have emerged, and although the findings are frequently controversial they are none the less believable for that.

For a hundred years or so, orchids were considered to be most closely related to another monocotyledon plant group – the lilies. But in the last year or two, DNA analysis, using a technique to compare sequences from different species, has shown that it is not the lilies that are most closely related to orchids, but a different group entirely, the yellow star-grasses. Although this is not entirely surprising, and perhaps not as remarkable as some of the relationships that have been revealed using DNA, it is significant because in taxonomic works plant families are grouped in a distinct order, based on their perceived relationships.

The discovery of antibiotics

Probably one of the smallest morphological differences that can be defined between organisms and yet one with the most profound implications for humans is the difference between antibiotic sensitivity and resistance in bacteria. We had, in the second half of the twentieth century, begun to believe that the scourge of infectious disease may no longer a problem for humanity, and indeed in some cases this is true.

For example the final extinction in 1977 of smallpox in the wild was a major triumph for the persistence of authorities in vaccinating huge numbers of people. You will notice the qualification of the word 'extinction' with the phrase 'in the wild'. There are several known sites, generally research institutes, that still retain cultures of this hideous disease. It is important to remember that the potential of these cultures for use in biological warfare is staggering, especially since in the case of an epidemic it would be quite impossible to produce vaccine in sufficient quantities to defend a population.

The last case of smallpox was found in Somalia in 1977, but only ten years earlier, in 1967, more than 118,000 cases were officially reported. Sometimes, when this particularly devastating disease has arrived on new shores, records of the progress of the disease have been kept. One such example of this is when, in 1962, a man from Pakistan arrived in London and travelled on to Wales. He became ill with smallpox and was admitted to hospital, but not before the disease had started to spread. By the time diagnosis was made and vaccination had brought it under control there were at least 50 confirmed cases and eighteen deaths. How much more alarming, then, to consider the possibility of there being no treatment to stop such a microbial disease, whether it is viral or bacterial.

The search for methods to control bacterial disease had been going on for many decades when the most famous of all antibiotics, penicillin, was first used as a medicine in the 1940s. The results were of enormous and instant benefit, although this was not the first antibacterial agent to be found. The first antibiotics were sulphonamides, discovered in 1932 by the bacteriologist and pathologist Gerhard Domagk (1895–1964), working at the German company I.G. Farben. The company was involved in a programme of laboriously screening hundreds of dyes for antimicrobial activity in the search for new drugs. The first one they found to be active worked well *in vivo* but not *in vitro*. The mystery was solved when it was discovered that the activity was a result of the body snipping off a chemical side-chain, thereby turning an inert dye into an active drug. At that time Domagk's daughter had a life-threatening infection and he was moved by desperation to try this red dye as a medicine for her. To his relief it worked and he was later rewarded with a Nobel Prize for his discovery.

The discovery of these early 'antibiotics' gave great impetus to the search for methods to kill disease organisms selectively. Unfortunately, however, the success of sulphonamides, was short-lived; resistant bacteria soon appeared because of the widespread and indiscriminate use of the drug and sulphonamides were no longer considered useful in treatment of disease.

Antibiotics as we now know them were first discovered at about the same time as sulphonamides, when penicillin was observed at work in 1928 by Alexander Fleming. The very first case upon which penicillin was used illustrates how seemingly trivial accidents can become serious problems if they are not treated. This first patient was a policeman who had cut himself, very slightly, while shaving. The wound became infected with bacteria called staphylococci, and they spread into his bloodstream causing potentially lethal blood poisoning. In an effort to save his life, the doctors decided to try penicillin, even though it was only available in tiny amounts. At first it seemed as though the treatment was working and the patient was improving, but even though the doctors attempted to increase the amount of penicillin by extracting it and recycling it from the patient's urine, there was just not enough and the patient died.

This was also the fate of the second patient on whom penicillin was tried, but the third, a fifteen-year-old who had an infected wound from an operation, recovered. The implications were obvious and the development of new antibiotics became the dream of the fledgling pharmaceutical industry.

Bacterial resistance

Soon after the introduction of antibiotics it began to be asked just how they worked. Antibiotics are really nothing more than just selective poisons, or toxins. They have a specific mechanism of action, but the first thing they have to do is get inside the bacterial cell. Once inside the cell the way in which an antibiotic inhibits the microbe depends upon which type of antibiotic it is. Penicillin, for example, blocks normal cell wall production so that the bacteria cannot multiply, while chloramphenicol interferes with the production of specific bacterial proteins.

The enormous success of antibiotics now makes us feel that infectious diseases are no longer a problem. Unfortunately they are, and are becoming progressively more so because of the indiscriminate use of antibiotics, especially as feed supplements in animal foodstuffs. Use of penicillin has become so widespread, being measured in tons per year, that we have been unknowingly carrying out our own experiment in natural selection with the human race as the test environment.

It is an unfortunate fact that a little more emphasis on understanding the genetics of evolution and natural selection in our medical schools might have reduced this problem to one of mere academic curiosity. For biologists in general and geneticists in particular, it has always seemed ridiculous to imagine that pouring vast amounts of antibiotics into the environment will not result in selection for resistance. These antibiotics are often used totally inappropriately, being prescribed as protection against viral infections, for example, against which they are ineffective.

Just as antibiotics work in a specific manner, so too does

antibiotic resistance. Resistance normally comes in the form of plasmids. These are small, circular pieces of DNA that are independent of the bacterial chromosomes and that can self-replicate within the cell to produce several copies. Plasmids contain enough genetic information to code for the production of enzymes that break down antibiotics. Because of the huge variety of antibiotics around, plasmids are usually specific to a particular antibiotic and each one can only produce an enzyme designed to break down one particular antibiotic. This is not always the case though; some plasmids can confer resistance to two widely dissimilar antibiotics. For example, the rather unexcitingly named bacterial plasmid pACYC184 contains the gene for chloramphenicol resistance as well as the gene for tetracycline resistance. Plasmid size is of interest here, because in contrast to the 3 billion base pairs in the human genome, this plasmid, a fairly average sized one, neither small or large, contains only 4,244 base pairs.

It is possible that these antibiotic resistance plasmids originated not in bacteria, but in viruses that were able to swap genetic material with bacteria. What is even more remarkable is the fact that it has been found that transmission of plasmids between bacteria involves something that was for a long time thought to be impossible: swapping of information between asexually reproducing organisms, in this case bacteria.

Bacteria swap genetic information in the form of plasmids via a process called conjugation. This involves the formation of a hollow tube to bridge the gap between two bacterial cells, through which the genetic information can pass. The material passing along the tube between the cells will include copies of plasmids going in either direction. This means that with the exchange of a plasmid, the recipient attains the ability to produce the enzyme to break down a specific antibiotic. Excessive use of antibiotics, therefore, creates a considerable selective pressure in favour of bacteria that have the relevant plasmid and can thus metabolize the antibiotic. This pressure results in populations of bacteria consisting entirely of cells that have this ability.

It was a long time before the nature of antibiotic resistance was understood, because although the conjugation of bacteria had been observed, the implications had not been fully understood. Plasmids have since been sequenced and susceptibility to antibiotics put to good use as a marker in screening protocols for other genes. In order to do this the DNA of interest is introduced into a plasmid which already contains the antibiotic resistance gene. If the cells are now grown on a medium that contains the antibiotic, only the antibiotic-resistant bacteria will grow. It follows from this that only cells that are antibiotic resistant and therefore also contain the gene under study will grow. The culture of such cells can result in huge amounts of DNA, suitable for sequencing the gene.

Nevertheless this is, perhaps, a very small use of limited value for plasmids; the fact remains that infectious diseases are not finished and still kill countless people every year. The initial enthusiasm for antibiotics in the mid-twentieth century, with the resultant inappropriate and overuse of them, has resulted in a resurgence of virulent bacteria.

DNA in phylogenetics

Because between closely related species nearly all DNA is shared in common, there are relatively few genes that can be used as markers of species. For example, nobody would dispute, I think, that chimpanzees and humans are different species. Zoologically they are not even in the same genus, let alone species. Nevertheless, we share all but about 1.3% of our DNA in common with chimpanzees. This means that DNA alone is very difficult to use as a method of delineating a species. Of course, as the evolutionary distance between species increases, so too does the variation in DNA, but for closely related groups the differences can be really quite small.

The more normal process when using DNA in phylogenetics is to have a recognizable group, usually defined by standard taxonomy, and then investigate the relationships within the group. This can give us considerable insights into evolutionary processes which can otherwise only be guessed at. Recently this has been used to demonstrate that the long-held belief amongst zoologists that the most primitive group of living reptiles are the turtles is probably incorrect. This belief was based partly on the morphology of the skull, which resembles far more ancient reptilian skulls from species long since extinct. But comparing DNA sequences from the subcellular organelles called mitochondria, it was possible to show that the closest living relatives are the crocodiles and birds, with lizards and snakes being more distantly related. So the shape of the skull must have evolved for a specific purpose which by chance resembles a more primitive type.

Zoo blots and comparisons between species

When DNA was first looked at closely using the process known as zoo blotting, an intriguing feature emerged. It seemed that the closer two species were related, the more they shared common sequences of DNA. This meant that DNA analysis could be used to establish how closely species are related.

Broadly speaking, a zoo blot is a method used to discover the degree of relatedness between different species. In a way it is very similar to the methods used to produce a DNA profile for individual identification, only instead of reacting a DNA probe with DNA from an individual, DNA from one species is cross-reacted with DNA from another species. Using enzymes to cut DNA produces fragments of different lengths which can be separated by running them through a gel under the influence of an electric current; the smaller pieces travel further than the larger ones. Once the DNA from a species has been treated in this way, a second species is used for comparison, digesting the DNA sample from species number two using the same enzymes. When put together under suitable conditions of heat and chemistry the DNA sequences from the two different species will hybridize, that is, they will bind together where the two samples of DNA match. The level to which hybridization occurs indicates the level of relatedness; if species are very close they will, logically, share much of their DNA in common.

As can be readily appreciated, the interpretation of the results can be a very intricate process. Understanding what these blots tell us can be very difficult because it is not just how closely related species are that affects the outcome but also the conditions under which the experiment is carried out. If the conditions of both chemistry and temperature are inadequately controlled it is possible to end up with a result which is patently absurd, so great care is needed for this kind of work.

This form of inter-specific comparison was originally developed using protein sequences rather than DNA, but it soon became clear that some proteins do not vary very much from species to species. This was not so much a surprise as confirmation of both the relatedness of species and the universal nature of life on this planet. But it did allow us to conclude that some proteins are not only essential for life but have virtually no room for variation because of this.

One such relatively inflexible protein is the mammalian hormone insulin, in which variation between species is relatively small. Presumably this reflects both the fundamental nature of insulin and its physiological activity in mammals. Insulin is a hormone produced by the pancreas, and is found in all mammals. If we accept that a common activity requires a common protein structure, then it is reasonable that the gene coding for it will, to a greater or lesser degree, also have a common sequence. In this case the insulin molecule has a level of activity which does not entirely depend on the exact sequence of amino acids which it contains, so it does have some scope for variation between species. Consequently it is not only in the amino acid sequence that variation occurs, of course, but also in the DNA sequence which determines the amino acids.

In contrast to this it has been discovered that some proteins are of such a basic importance to life that there is virtually no room for variation. One such is an apparently obscure protein called histone H1, which is fundamental to the process of folding DNA into chromosomes. Wherever DNA is found, so too is is the histone H1 protein. And not only is it ubiquitous, its structure is also almost completely inflexible; in fact there is virtually no difference between the amino acid sequence of histone H1 in peas and that in humans.

In reality this inflexibility is not so surprising. It is the function of the protein/DNA complex to construct a regulated environment for gene expression via the use of a chromosome. Since folding and packing of DNA is essential and the method of folding fairly limited, it seems reasonable to expect that since DNA is universal the folding proteins will be consistent in structure as well. What is more of a surprise, though, is that there is so little variation possible in this protein across the huge range of life that we have on this planet. Knowing this, it should be possible to locate the histone genes quite easily and to trace them from one species to another, bearing in mind the ability of the last base of the triplet code to vary without affecting the resultant protein very much.

We can see from this that there is a unity to life and the manner in which it is constructed and controlled. Such uniformity in the very basics of control of DNA and therefore life is in itself remarkable, but it also gives us insight into the way that life has evolved on earth. It does seem reasonable that given a consistent method of inheritance, DNA, and the consistent nature of so many gene products, proteins, life had a common origin. So while early taxonomists had a desire to try to relate extant species to each other, they had no idea that their phylogenies reflected the true relationship through time of one species to another.

Comprehension of evolutionary time

Early taxonomists tended to limit themselves to the naming and grouping of species without reference to the relationships of one group to another through time. In this they were constrained by the work of Archbishop Ussher. It was he who, in the years 1650–54, set the origin of the world as 23 October 4004 BC. This calculation was necessary for humanity to explain its own existence within the framework of the Bible. After all, an intelligent and curious humanity will inevitably result in questions being asked, but if no rational explanation is available it is the nature of humans to make one up. We can appreciate that the awesome notion of a planet unimaginably old would be impossible to understand in the seventeenth century.

Even today it is difficult to comprehend the age of the planet, comprehension being quite different to merely saying the numbers. For example life, as far as we can estimate, is about 3,500 million years old, maybe older. This is an impossibly huge number to visualize. To try to put this into context, writing is a matter of a few thousand years old and Christianity only 2,000 years old. Our comprehension and perceptions of time are limited by our close horizons, which for most people do not extend beyond a single century. It was natural, therefore, to put the world and all life into a human perspective of time, resulting in the creation myth. This not only gave us a distinct starting point for the planet, but also for all the life around us, which apparently arose fully formed. Through this we also gained the notion that we had dominion over life.

Today we have a better idea of how things are related to each other, but let us not be arrogant and think that we have the entire story. It is still difficult for most people to understand that we do not control the planet, and yet in many ways we have no more control over our collective destiny than a pigeon does.

Southern blotting and advances in DNA research

When DNA was first described as the genetic material it was immediately realized that it contained an enormous amount of information of use in the battle against disease. But even though it was seized upon for the potential information it offered, it took 25 years before a technique was developed that could be used to analyse DNA in sufficient detail to be of use. This technique was the Southern blot, described in Chapter 3, in which fragments of DNA separated on a gel using an electric current can be fixed on a membrane. Southern blotting opened up the possibilities for research on DNA and, along with the polymerase chain reaction (PCR), has had considerable impact on our understanding of animal relationships and evolution.

Some forms of DNA, most specifically mitochondrial DNA, can last for very long periods under suitable conditions. So how can we use this knowledge to help us understand the natural world? With a little ingenuity it is possible, bearing in mind that some genes are highly conserved between different species, to compare DNA sequences between different species. If we assume that mutation rates tend to be constant over long periods of time, which is more or less correct, then the degree of difference between the sequences of genes of different species gives us a measure of the relative time that has passed since the species diverged.

The fact that some genes are conserved over extraordinarily long periods of time makes it possible to take their products out of one organism and put them to work on material from an entirely unrelated species. An example of this is the use of DNA polymerase from bacteria to replicate DNA from any organism, including humans, in a test tube. This has made it possible to design systems which can replicate DNA artificially from very small samples. This process takes advantage of the universality of DNA and the fact that the enzymes that replicate DNA in nature are all very similar in all species of plant and animal. Thus a bacterial enzyme can be used to artificially replicate DNA in the laboratory. This is the process called the polymerase chain reaction, abbreviated to PCR. PCR replication is exponential; in theory it is possible to produce enough DNA to analyse from a single molecule.

Once replicated, a sample of DNA that was previously too small to analyse becomes eminently suitable for experiment. In the past it had always been assumed that if DNA could not be detected on a Southern blot then there was not enough there to do anything with. This meant that although it was always possible to go through the processes of DNA extraction, it was not always possible to show that DNA had actually been extracted from the original material if there was very little present.

When, in 1983, Kary Mullis devised PCR the rules changed. All of a sudden the definition of presence or absence of DNA revolved around whether PCR could produce a product. PCR made it a practical proposition to look at all manner of old DNA. This technology has been used on ancient human remains as well as extinct species of animal.

Quaggas and Jurassic Park

Perhaps one of the most interesting pieces of such work has been on DNA extracted from an extinct species of wild zebra called the quagga. This extraordinary animal was quite unlike any other zebra, although occasionally Burchells' zebras with aberrant markings were mistaken for it. Quaggas were more or less brown, with white legs and tails, but with zebra stripes just on the neck and head. They were very common throughout southern Africa until the nineteenth century, when invading humans hunted the species to extinction. It is recorded, to the shame of humankind, that the last wild quagga was killed in 1878 and the last specimen in captivity perished alone in an Amsterdam zoo in 1883.

Because the quagga is so recent an extinction of what was a common animal, there are relatively large numbers of skins about. When an animal is killed and skinned the cellular material still retains a surprisingly large amount of DNA, preserved by the same process as preserves the skin. This is more or less the same outcome as when an Egyptian Pharaoh is mummified but is quite different to the chemical processes used to tan leather. Tanned skins are quite useless for DNA analysis but dried or preserved skins and other body parts are usually fruitful sources of DNA samples. Indeed, ancient Egyptian mummies have been investigated in attempts to find out if they suffered from any inherited diseases and how they were related to each other.

In the case of the quagga the preservation of the skins meant that there was plenty of DNA available for analysis. But let me pre-empt the next step in your train of thought. You may imagine that if the DNA is still around then it may be possible to recreate the whole animal, along the lines of the film *Jurassic Park*. Sadly, no, the fact is that extinction is indeed for ever and it will never be possible to recreate an extinct animal from fragments of its DNA. The scenario outlined by Michael Crichton in *Jurassic Park* is a very clever and fascinating work of fiction, but it will always remain just that – fiction.

In the case of dinosaurs there are some very practical reasons for the impossibility of recreation. First, it is now thought that the original experiments in which it was claimed that DNA could be extracted from a termite trapped in amber for 25 million years were a mistake. Painstaking work over many years has not shown any ancient DNA surviving in insects trapped in amber. Secondly, although very small samples can be analysed using PCR, the process is dependent on the presence of primers, even ones that work randomly, and the production of these is itself dependent on either knowing, or assuming, something about the target sequence being investigated. This is knowledge which is simply not available about dinosaurs. Besides this, one of the reasons PCR works so well on old DNA is that it only replicates very short sections, one copy of which is nearly always found intact somewhere in the extracted DNA. An alternative approach would be to find a source of DNA which provides enough material for sequencing. The possibility of this is remote.

Of course, even if enough DNA were found and it was certain that it came from only a single species, the problems would not be over. Just knowing a DNA sequence is not enough, even the whole genome complete with control sequences and all the bits and pieces that switch genes on and off. There are, as we have discussed, proteins associated with DNA that control gene expression and activity during cell growth and development. These would need to be present. Another aspect is that DNA, however well preserved, does break up with time. The decay caused by bacterial and fungal enzymes damages the sequence by introducing breaks in the DNA double helix. Even if the whole sequence is known, if the order is incorrect then the knowledge would be of limited value. By analogy, a dictionary may contain all the words found in Shakespeare's plays, but if you do not put them in the right order you do not have *Hamlet*. In the case of DNA, if the order is wrong you end up with abnormal development and tumour growth. Genetic activity is probably one of the most tightly controlled biological functions imaginable; if it goes wrong, everything goes wrong.

So what is the additional information we need to make an animal? Simple, the order of genes down a chromosome, not just the DNA sequence of the genes, but the order of the genes themselves. Without that all we have is a nonsensical listing of data. Just as a random collection of names and telephone numbers on its own means nothing, and it is only when we sort them for use into a telephone directory that it gains some value, so the DNA sequence on its own has no meaning. This is not a small point, the magic of information is what it allows you to understand and do; simple ownership of a list is of no value.

It is this additional data which is lacking in these cases. But as well as this it must always be remembered that analysis of very small samples will almost inevitably leave gaps of unknown sequences. These gaps might even contain vital pieces of information without which no animal can be raised. Without a complete DNA sequence, in the right order, you can be sure that embryology will be disrupted.

It is a very delicate process to raise an entire organism from a single cell to independence of function. It involves the right environment and the right DNA, switching genes on and off in the right places at the right times. A wrong switch during growth will result in disrupted development. A wrong switch after growth and the result is a tumour. The very powerful and highly conserved genes that control growth and development are equally good at killing the organism if they are turned on at the wrong time.

Arguments for and against the possibility of resurrecting a long extinct species from ancient remains inevitably wax and wane. There is always a fancy that science will eventually be able to do anything we can imagine, like space travel to distant suns and planets, or communicating with alien species. But this particular task may be just beyond our ability. In the long run it may be better to go back to basics and try to breed back from one of their descendent species.

Alien life forms

During the twentieth century, especially in the second half, perceptions of humanity have altered. Not so very long ago there were still areas on the earth unexplored and unseen by westerners. This inspired many works of fiction, such as *The Lost World* by Sir Arthur Conan Doyle. As we became more familiar with our surroundings, especially through the agents of television and film, the scope for science fiction based purely on this planet has become less. Alternatively, I suppose it could be claimed that our horizons have widened with our imaginations, so we look to the skies and the hope of alien intelligence for our science fiction entertainment. So what can terrestrial genetics have to say about this? Probably a great deal.

Consider the vast range of shapes and sizes throughout the animal and plant worlds. Not just mammals, but of bacteria, plants, insects, molluscs, birds, fishes, all the vast myriad forms of life. These all share nucleic acid as their genetic material, a fact that attests to a common origin of life on this planet. As discussed earlier, the discovery that enzymes from bacteria can be used to artificially replicate DNA from animals and plants is an awesome demonstration of the universal nature of life on earth. But what of other planets? It is little short of hubris that modern film makers nearly always portray alien species as two-legged, bilaterally symmetrical humanoids about two metres tall. There is no reason to believe that aliens should be like this, even assuming that there really are beings on other planets, which we do not know. I see this as almost like a form of colonialism; we are dominant and therefore we make them

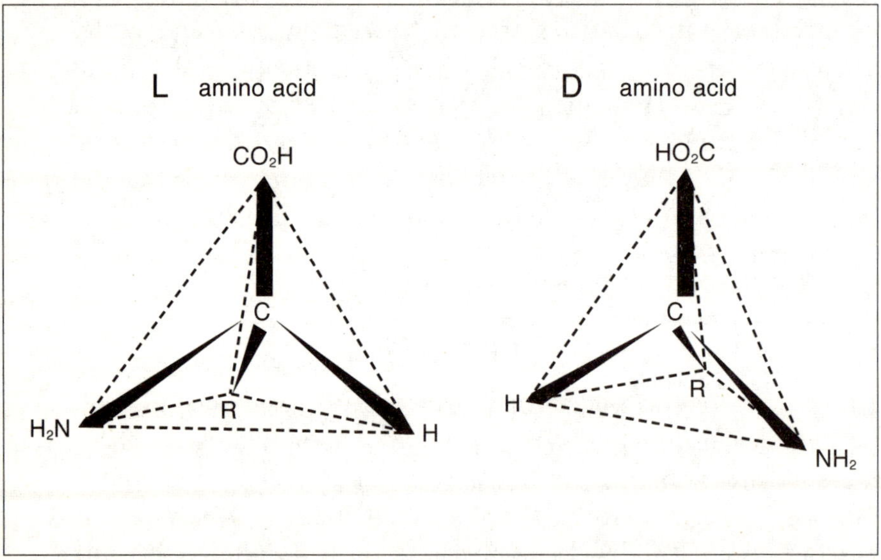

Different forms of amino acid. Amino acids come in two types called D and L forms. D and L amino acids are chemically identical but they are mirror images of each other. Biological organisms only use the L forms.

in our image. Why should alien life forms not appear as the Super Intelligent Shade of the Colour Blue as described by Douglas Adams? It is just as likely, or unlikely.

Similarly, there are some films, TV series and books that recount tales of interbreeding between alien species, resulting in jolly hybrid crosses. Not a chance. Crosses between complex organisms on this planet are virtually impossible, and we share a common genetic material. In fact, the imbalances resulting from interspecies matings are devastating, and even when it does work, as in a donkey and horse cross or a lion and tiger cross, the resulting animal is almost invariably sterile. Just as the assumption that alien species look like us and communicate by sound and light is totally illogical, so is the idea that alien species will have DNA as their genetic material. And if they do not use DNA then exchange of genetic material, which for very practical reasons may not be possible without artificial help, would be impossible. We can relax in the knowledge that formation of a viable hybrid is out of the question.

There are also very basic biochemical reasons why it seems unlikely that we could interact with alien species in other ways, for example in sharing or, heaven forbid, providing a source of food. Amino acids, the basic building blocks of proteins, come in two different structural forms called isomers, and are either L or D forms, as illustrated on page 108. These are mirror images of each other. Now, life on earth is dependent entirely on the L form of amino acids, but there is nothing to say that even if alien planets had a biochemistry based on proteins, they would not be constructed of the D form of amino acids. This would make our sources of food totally incompatible: they could not eat our food and we could not eat theirs. Although this would have the advantage that they could not perceive us as a tasty morsel, it would make it very difficult to coexist.

Of one thing you can be certain, if alien intelligence is ever discovered, you will be more closely related to a tree in the park than to any alien species. You can also be sure that reality will be far stranger than anything we can imagine.

FIVE

When the system fails

DNA analysis in the diagnosis and treatment of disease

Although it is often said that there are about 100,000 genes in the human genome, this is probably an overestimate. If we take a more conservative estimate of 60,000 to 70,000 genes in the human body, we still know of more than 4,000 which can become impaired or non-functional. These may be severely impaired, but are still just about compatible with life. We know this because, when they go wrong, the result is often, though not always, a disease or disorder.

It is quite likely that every gene can be altered in some way; after all, modification and mutation of genes is the driving force behind evolution, but these 4,000 clinically significant mutations are the ones we know about. Of course, as described earlier, a gene does not necessarily become defective if it picks up a mutation from radiation or a chemical assault. This is because a certain amount of variation in DNA sequence is tolerated without altering the final protein product. In the same way, some variation in the amino acid sequence of a protein can be tolerated without changing the ability of the protein to carry out its designated function.

Although it is hard to quantify such figures because of the difficulties in diagnosis, broadly speaking, genetic defects in humans, not counting gross chromosomal conditions, affect about 1.25% of all live births. So the risk is both tangible and real. Of course, some mutations result in very mild conditions that hardly affect the individual and some are so devastating that they are not compatible with life; these latter will not be seen and therefore do not appear in our list of 4,000 mutatable genes. But it is clear that with such a high rate of aberrations, a great many people are faced with the problems of dealing with the results of genetic changes which result in disease, on a day-to-day basis.

From mutation to disorder

If mutations occur with equal frequency in any given length of DNA, which is what we would expect from the simple observation that there is no long-term pattern in the sequence of DNA, then we would expect small genes to have less mutations than large ones. But this does not accord with what is observed: some genes seem to be particularly susceptible to rearrangements.

Of course, some genes are so important that any mutation is devastating, resulting in damage creating a completely non-functioning cell. We are not talking here about stunted growth, but about cessation of development. In humans and other complex animals this might take place at such an early stage that implantation of the embryo would not even occur. Indeed, in work with cloned embryos it has been noticed that the successful rate of development is far less than would be expected if they were all identical. Although no explanation has been produced for this, it is likely that it is associated with genetic problems.

The occurrence of certain very similar stretches of DNA sequence in unrelated species indicates the fundamental importance of the proteins coded for by these genes. Very highly conserved DNA sequences are usually associated with genes that have little or no ability to alter without rendering them useless. A tiny alteration in the gene, even of only a single base, might easily stop development almost before it has started.

It is not, of course, the DNA itself which causes problems when altered, it is the gene product, the protein, which is the final culprit. Proteins are always the first product of translation of a gene, whether they are in the form of enzymes or structural proteins. Enzymes are used by the cell to modify all the chemical functions of an organism, from creating sugars from carbon dioxide and water in plants during photosynthesis, to the metabolism of these same sugars for energy in animals.

The idea that defective proteins are entirely responsible is not the complete story, however, as we know that the delicate balance between different, normal, proteins is also important. The maintenance of a balance between gene products is also essential; that is, the quantity of one protein relative to another. This is why the existence of three copies of a chromosome, rather than the normal two, can have such devastating effects on an individual, simply as a result of imbalance of gene products. Cases of trisomy, as this is called, include Down's syndrome in humans, which occurs in people with an extra chromosome 21. Because of our complexity there are very few trisomies involving complete chromosomes that are compatible with life and those that do exist involve relatively small chromosomes. But all of them have very serious consequences for the patient and his or her family.

Deletions and duplications

Sometimes, imbalances caused by deletions or duplications of large sections of a chromosome also create severe medical problems. Some of these chromosomal imbalances are so large that they are visible using a light microscope. Nowadays pictures of chromosomes are commonplace and diagnosis of these conditions has become routine. It was very soon after it was established that the normal human chromosome number is 46 that the first of these conditions was described.

There are many ways in which deletions and duplications can result in difficulties for the individual. First, there is the problem of protein imbalance, as with trisomies, although as we shall see, not all proteins need to be in perfect balance. But deletions have another trick up their sleeve which is potentially far more devastating. We all carry defective genes but we also carry two copies of every gene, so if one is a dud, the other one can still produce the protein necessary for the maintenance of life. This is, in effect, what happens when a person is referred to as a carrier of a recessive disorder. But if a deletion removes a normal gene from the panoply of information, it may 'uncover' a defective gene that is incapable of producing a functioning product. This can result in a cell, or even a complete person, that has an enzyme missing. Assuming that it is not so severe as to cause death in early development, the problem may only manifest itself later in life. Such conditions are known, and because our enzymes and proteins are so fundamental to normal metabolism, their absence will inevitably be deleterious.

Spotting single gene defects rather than gross deletions is, as can be imagined, considerably more problematic. Genes on a chromosome are far too small to be seen individually. It is only when they are packed into large blocks, in the form of chromosomes, that they can be seen to be present, absent or inverted. Without the ability to see individual genes it is only by using chemical analysis that it is possible to determine exactly what is going on. And until recently it was only the protein product that could be analysed, not the gene itself.

A quote from R.C. Lewontin of Harvard University is worth repeating here. He said that 'there is not a single case where knowing a gene sequence has led to a treatment for a disease'. This is true so far, but probably not for much longer; certainly in the next few years we can expect at least one attempt at gene therapy to bear fruit. In the mean time, knowing a sequence does help in our understanding of genetic disease, and that in itself is a positive step in the right direction.

It is inevitable in science that the more we understand, the more we are able to use and manipulate what we know. This might be perceived as not altogether a bad thing when it results in us being able to manipulate our personal environment to make it more comfortable. But when manipulating our environment causes us or our neighbours problems, we have to tread carefully when using our knowledge.

Polygenic diseases

Broadly speaking there are two groups of genetic disease other than gross chromosomal changes: single gene defects and polygenic diseases. Single gene defects are really self-explanatory; some of the best known are familiar diseases, such as cystic fibrosis and muscular dystrophy. Polygenic diseases can be thought of as having variable numbers of genes involved. They form a rather odd group of conditions, with nothing in common except that they are caused by groups of genes, rather than just one. These diseases are not the product of a single dysfunctional gene, but of many different genes going wrong and the results are often subtle and difficult to quantify. Looked at from another angle, the degree of expression of a polygenic disease depends, at least in part, on the number of changed genes in the set. I use the word 'changed' here rather than 'abnormal' because the protein product works perfectly well, it is not in that respect abnormal, but in association with other gene products the result may be detrimental. It is also worth remembering that the genes involved in a polygenic disease do not need to be related by a common metabolic pathway, it is the accumulated result of the gene products which causes the problem, not simply a protein that cannot do the job for which it is designed.

Some polygenic conditions can be modified by the actions of the person affected. For example, in families with a history of heart disease, modifying the diet may well improve the potential outcome to such a level that the problem will seem to have disappeared. It has not actually, of course, because the genes responsible are still there, but to all intents and purposes the problem no longer exists. As we shall see when we look at ethical questions of genetics and health or life insurance, it is polygenic diseases which are likely to cause the greatest disputes with the insurance industry.

As we have seen, polygenic systems are quite flexible in their results; in interaction with the environment, the outcome for the individual will be a mixture of both genetics and environment. This is after all what life is all about – the interactions of inherited characters and environment. For example, humans are not well adapted to cold climates, our genes have seen to that, but we live there by wearing clothes for warmth. Although, unlike the polar bears, we are genetically poorly equipped to survive in extreme cold, we modify our environment with warm clothes so that our survival is ensured.

Inherited conditions

It is not always necessary to know which gene is responsible for a disease, nor even the biochemical pathways involved to be able to see that a disease is genetic in origin. In fact sometimes it is easy to recognize that a condition is caused by a defective gene while knowing absolutely nothing about the mechanism involved. While infectious diseases tend to appear sporadically, the commoner genetic disorders follow a pattern of inheritance that rapidly makes it obvious that the disease runs in families.

Here again the picture becomes a little more complicated as we delve deeper. Most genetic traits, not just genetic diseases, can be thought of as either recessive or dominant. All genes, except those on the sex-determining chromosomes, X and Y, are present in two versions, sometimes the same and sometimes different. One of these is inherited from the father and one from the mother. The two versions of the gene may be sufficiently different to end up with a different final result. For example, if you look at people's ears they can generally be put into one of two categories, depending on the way in which the earlobe is formed. The two variations of earlobes are called attached and free, depending on the way the lobe is attached to the side of the head. The type of earlobe is inherited in a quite straightforward manner and seems to be controlled by a single gene. This means that we can say that 'attached' is dominant and 'free' is recessive. Exactly how the mechanism operates is obscure, but it really is quite remarkable when you think about it. What could possibly be the reason for this seemingly pointless difference in the population?

Another similarly strange and pointless difference can be found in the ability to tongue roll. Either you can roll your tongue into a tube, or you cannot; the ability is controlled entirely by your genes.

When it comes to looking at diseases this difference between dominant and recessive can be very important. If a disease is inherited as a dominant characteristic, then anyone carrying the faulty gene will be affected. Huntington's chorea is an example of a disease of this type. On the other hand, if the disease is inherited as a recessive characteristic then only people with two abnormal genes will be affected; those with only one defective gene will be perfectly normal, but termed carriers. Cystic fibrosis and muscular dystrophy are both recessive diseases.

Sex-linked disorders

Some genes are only found on the X chromosome and have no counterpart anywhere else in the genome. Because women have two X chromosomes, a defective gene on the X chromosome can be carried by a woman without affecting her adversely. She would be what is known as a carrier. In the next generation, each of her sons would inherit one X chromosome – either the defective one or the normal one – from her and a Y chromosome from their father. Those unfortunate enough to inherit the defective X chromosome would be affected by the disorder. Without the back-up of a second X chromosome, the male offspring have only the one X chromosome to rely on and if that is defective then the disorder will be apparent, sometimes with devastating consequences. In the female line, half of the woman's daughters would be carriers of the defective X chromosome and half would be unaffected, having two normal chromosomes.

As can be appreciated, members of families known to have one of these sex-linked disorders can be faced with very difficult decisions if no test for carrier status is available. One such condition is haemophilia, a disorder that has been well documented in at least one extended family over several generations.

Haemophilia and the royal families of Europe

Haemophilia affects the mechanism the body uses to clot the blood. It is a sex-linked recessive disorder, which means that males can be affected and females can be unaffected carriers. In theory it would be possible for a girl to be affected if her father was a haemophiliac and her mother was a carrier, but survival would be most unlikely beyond puberty. It has been reported that in at least one case an affected woman survived childbirth and produced an affected son.

The major symptom of haemophilia is uncontrolled bleeding

following injury, leading to excessive bruising or loss of blood; in practice the injury may not be apparent, and sometimes bleeding can be spontaneous. The bleeding takes the form of slow oozing, out of all proportion to the injury and sometimes lasting for weeks at a time, resulting in serious anaemia. Very often cases first present themselves early in life because of recurrent bleeding into the joints or unexplained bruising.

Strangely enough, although blood clotting is always impaired in haemophilia, the causal gene is not always the same. Two forms of haemophilia have been identified: types A and B. If blood from a person with haemophilia A is mixed with blood from someone with haemophilia B, the resulting mixture is found to clot normally. The assumption is that the two forms have a different component missing and when mixed together all the necessary material to clot properly is present. This was confirmed when it was determined that haemophilia A lacks blood protein factor VIII, while haemophilia B lacks factor IX. Haemophilia B is sometimes called Christmas disease after the first patient described with it.

Even though genetics is a relatively new area of study compared with other sciences, the recognition that haemophilia occurs in families has a long history. Indeed, the Talmud, as written down between the second and sixth centuries AD, has a provision for dispensing with circumcision of males when an older brother has a tendency towards excessive bleeding. In the early nineteenth century the form of familial inheritance of haemophilia was first described in three unrelated families from New England, but it was not until 1911 that haemophilia was shown to be consistent with inheritance on the X chromosome.

Perhaps the most famous group of people afflicted with haemophilia are the members of the royal families of Europe. In these families we know not only the nature of the defective gene, but also when the problem first manifested itself because of the complete and unambiguous records kept regarding parentage in royal families. It is now known that within the royal family the mutation causing haemophilia first appeared in Queen Victoria. It is probably just a coincidence that her father, Edward Duke of Kent, is known to have carried a defective gene for another, unrelated inherited disease, known as porphyria. The existence of this disease, which affects the breakdown of haemoglobin, could be traced back in the family as far as the sixteenth century and Mary Queen of Scots. It may be thought that, although Edward did not pass the porphyria to his daughter Victoria, the appearance of two inherited disorders within one family was significant. However, we must bear in mind that detailed genealogical family histories such as are available for the royal families of Europe are very rare indeed. If we had relevant details for other families and their ancestors the occurrence of two mutations within 350 years might not be found to be so unusual.

Knowing the precise origin of such a mutation and how it travelled through families is in itself a valuable insight, but not the only one from such studies. Haemophilia affected both the Russian royal family and the Spanish royal family. For the Russian royal family the problems started when Alexandra, grand-daughter of Victoria, married Tsar Nicholas II of Russia. Both Alexandra and her mother, Alice of Hesse, can be shown to have been carriers from analysis of the family pedigree. Although we do not know the carrier status of the four female children of the Tsar because of their assassination, we do know that the fifth child, a boy called Alexis Tsarevich, was affected and suffered considerably from his haemophilia. Similarly Victoria's youngest daughter, Beatrice, was a carrier and of her four children with Henry of Battenburg two of the boys were affected, one was normal, but her only daughter, Victoria Eugenie, was a carrier. When she married Alphonso XIII of Spain the resulting family had two affected boys, two normal boys and two normal, non-carrier, daughters.

The British royal family continued after Victoria with Edward VII, who was himself unaffected and therefore so, too, were his children. When Prince Philip, a distant relative of Victoria via Alice of Hesse, married Princess Elizabeth, it was possible that the defective genes could have returned to the royal family. Fortunately, this was not the case and the family is now free of haemophilia.

When these extended families are looked at it becomes clear that the statistical sample is just too small to produce the expected ratios of normal to carrier to affected. This is one of the reasons that prolific breeders, such as bacteria or fruit flies, make better experimental organisms for elucidating Mendel's laws.

Alcaptonuria and phenylketonuria – single enzyme defects

For some genetic conditions, effective treatment has been available for much longer than knowledge of DNA. Take, for example, the rare, recessive condition called alcaptonuria, a disease which causes dark brown discoloration of the skin and eyes and progressive damage to the joints. Alcaptonuria was first described in 1902 by a famous biologist, William Bateson, who, by the way, was also responsible for coining the word 'genetics' and became the first professor of genetics at Cambridge University. What Bateson recognized was that alcaptonuria was inherited as a recessive condition. But although the disease was described in some detail over the following years, these were functional descriptions of the condition, it remained unknown exactly what the biochemical problem was and, needless to say, nothing was known about the genetic basis of the disease.

In the end, the complicated biochemistry of alcaptonuria was slowly worked out on the basis of the end result, the symptoms which appear in the affected individual. It was found that the darkening of the cartilage at the joints and the tendency towards arthritis was entirely due to a single problem, the inability to break down 2,5 dihydroxyphenylacetic acid, often just called alcapton or homogentisic acid. The most noticeable and easiest diagnostic test is that urine from affected individuals turns black on exposure to air. This is due to homogentisic acid being excreted in the urine, which does not normally happen. This change in the colour of the urine had been noticed a long time in advance of the disease even being named; it is after all, a very obvious thing to notice.

The defect in metabolism causing alcaptonuria, finally analysed in the 1920s, was shown to be a deficiency in a single enzyme, stopping the body from oxidizing the alcapton in a controlled manner.

Today there are several known metabolic diseases of this type, all of which are based on an inability to correctly metabolize the amino acid phenylanaline. This results in accumulation in the body of either phenylanaline itself, or various metabolic breakdown products, depending on the exact nature of the genetic lesion. Some of these conditions can be treated relatively easily, although none of these treatments can be called a cure.

In the case of the condition called phenylketonuria, simply reducing the levels of phenylanaline in the diet can be very effective in controlling the symptoms. If this dietary control is started early enough, the symptoms, which include mental deficiency, epilepsy and behavioural problems, can be virtually eliminated.

People controlling a disease through modifying their diet are often helped by the clear labelling of foods by the food industry. You may well have seen products on sale labelled as containing phenylanaline, most often on soft drinks which appeal to children. If you know which products contain phenylanaline, then the diet can be modified accordingly. It is only manufactured foods which are labelled with this sort of information; there is little point in going to those limits with generic foods, such as green vegetables and meat. It is relatively easy to determine the composition of these foods. Within reasonable limits they will not alter significantly from season to season and tables of components are readily available. With manufactured foods, however, the content of all ingredients is under the control of the producer, and the shopper has to rely on the information supplied on the label.

Unfortunately, although the symptoms of diseases such as phenylketonuria and haemophilia can be ameliorated, there are still no cures for these conditions. This is the same for all genetic diseases, and for most genetic defects the only long-term solution is going to be gene therapy and that depends upon knowledge of DNA. Gene therapy requires real information and understanding about the fundamental genetic lesions and the DNA sequence of both the normal and abnormal genes.

Linkage analysis

So where does this take us in respect of genetic disease? Well, if we do not know the nature of the genetic lesion, we have no chance of curing the disease, and here I do not mean simple alleviation of symptoms as in phenylketonuria, but actually curing the disease. What we need to know is exactly what the genetic lesion is. Needless to say this is not always so easy to discover. The question is where do you start? We know that genes are carried on chromosomes, and we can see chromosomes, but not the genes, so how do we proceed?

There are so many questions and yet apparently so few answers, but there are methods which we can use to help us get to the bottom of the problem and help our understanding of genetic diseases. If we are going to attack a genetic disease the first thing we need to know, probably before anything else, is where in the human genome the gene for a particular condition is to be found.

It is not by chance that the seventeenth-century philosopher Sir Francis Bacon is quoted as saying that 'knowledge itself is power'. He fully appreciated that unless we understand what goes on, we not only know nothing, but we cannot take the next logical step in applying this knowledge to a practical problem. In our case, then, the first step is to track down a gene, to find where it is in the genome and possibly start teasing out the secrets it holds.

If someone said to you that you needed to find a bluebell wood, what would you do? Knowing that bluebells are associated with ancient woodlands, the first response would be to locate a woodland that was a likely habitat. Similarly, though with a rather less picturesque hunting ground, it is the same when hunting for human genes; you have to start from a point that you already know all about.

Starting the hunt for defective genes that cause disease begins by looking for distinctive DNA landmarks that are associated with the disorder being investigated, but that do not cause the problem themselves. These landmarks are called markers. There have been many different techniques developed in the past to do this, but the one method that has yielded consistent results is linkage analysis. I am not implying that linkage analysis is easy, far from it. Indeed, if you want a difficult and frustrating job, just try to locate a specific gene, whether responsible for a disease or not. This is probably the most taxing activity ever to confront a geneticist.

Linkage analysis relies on finding an association between a specific, known, piece of DNA and the defective gene. The degree of linkage depends on how close the two pieces of DNA are together. If they are next to each other then the linkage is regarded as tight; the further away the two sections of DNA are from each other the weaker the linkage is, until it is no longer of any diagnostic value.

In very small populations this phenomenon of linkage can work very much to the gene hunter's advantage. For example, the population of Iceland is ideal for linkage studies and it has been appreciated by their own scientists just what a valuable resource a unified group of people extending back to the ninth century can be. The population of Iceland is about 270,000, but it is not just the size of the population which is important. The tradition of very careful family record-keeping in Iceland means that records of health in Icelandic families can go back centuries. So accurate is this that in 1997 a local investigative team demonstrated that of about 1,200 asthmatics found on the island, 107 of them could trace their heritage back to an individual living eleven generations earlier. Whether this will be of practical use in tracing a genetic component for asthma remains to be seen, but we can be sure that there are many more than 107 descendants of that one individual, and they apparently do not have asthma. As it can be appreciated, such detailed and meticulous record-keeping is an excellent starting point for gene hunters looking for other disease genes.

Crossing-over and the search for markers

The reason that linkage works well only when the known and unknown sections of DNA are close together is because of an event that takes place during cell division called cross-over. When cells go through the process of reduction division, or more accurately, meiosis, to produce either sperm or eggs, two very special events occur. The first is that the number of chromosomes is reduced by half so that when fertilization takes place and two cells fuse, the complete number of chromosomes is restored. The second special event is called cross-over, a process of extreme importance for all sexually reproducing organisms. What happens during this process is that DNA in equal quantities, and it is very important that it is equal quantities, is swapped between the maternal and paternal chromosomes of the adult. By this method, when your germ line cells undergo meiosis, the chromosomes which you inherited in equal numbers from your parents shuffle bits of DNA between them. Because of this process of crossing-over it is virtually impossible for an individual to inherit an entire complement of DNA from a single grandparent.

Crossing-over is very well regulated and it rarely goes wrong. If it does go wrong, however, it results in unbalanced chromosomes, with one having more material than its counterpart. So how can we use this phenomenon of cross-over to help us locate a gene? Well, broadly speaking, the process of crossing-over occurs more or less uniformly down the length of a chromosome. Consequently the closer together two genes are located down a chromosome, the less likely it is that crossing-over will happen between the two. If two genes are very close together they are tightly linked and will tend to be inherited together.

If we can find a known, testable piece of DNA that always seems to travel with a gene which we do not know but are interested in, we have a method of telling whether the gene of interest has been inherited. If one of the two DNA fragments is a disease-causing agent, which we might not know, and the other is recognizable by some biochemical assay, by determining whether a person has the marker fragment we can gain some idea of the probability that he or she also carries the disease gene.

Of course, finding a linked marker in a family is a difficult and time-consuming activity that might never produce a result, which is why cooperation between scientists and populations such as the Icelanders is so helpful. Markers are often defined as areas of DNA which in one version may be susceptible to cutting with enzymes extracted from bacteria and in another form has lost the ability to be cut. Another type of marker which can be used is the size of a repeat sequence, a short tandem repeat, which can be measured accurately by using PCR. This can work very well in simple cases, where all that is required is to know whether a person is either a carrier, or perhaps affected.

One of the ultimate goals, therefore, is to identify the gene responsible for a disease so that it can be tested directly. Without this information no further work can be attempted, and this includes any chance of gene therapy. We need this information because the identification of carrier and potentially affected individuals by tracking markers in families still does not tell you what the ultimate biochemical problem is. Given enough information it is possible that treatment, even a cure might be possible. This latter outcome, gene therapy, is the ultimate goal of all genetic treatment. At this point genetic disease is no longer to be seen as something about which nothing can be done; a dark shadow on a family which is never spoken of. It is the point at which it becomes not just treatable, but curable.

The search for disease-causing genes

Those conditions which cause such distress in families by significantly shortening a life are bound to be the ones researched most vigorously. Besides, some of these very diseases also happen to be among the commonest. Cystic fibrosis and Duchenne muscular dystrophy are both relatively common inherited diseases and considerable research efforts have been and are being expended on the search for the genes responsible for these conditions.

To locate a specific gene to the point of being able to identify the very sequence of DNA involved in a particular disorder requires the use of very many markers and extended family studies. Sometimes this can take years and involve huge multinational teams. If there was ever an area of science which is the antithesis of nationalistic activity, it is the incredible cooperation shown between different groups looking for genes involved in diseases.

Huntington's chorea was tackled like this. Large research groups in many different countries shared information freely. This was in part due to the massive resources held in common, like the large panel of tissue culture samples from multigeneration families that are kept at the Centre d'Etude du Polymorphisme Humain in Paris, usually shortened to CEPH. Another huge multinational undertaking is the Human Genome Mapping Project (HGMP), which coordinates all the sequence data from laboratories all over the world and allows access to any scientist with a need for the information to help in their research.

Diagnosis of genetic disorders

The next question is, if we have the DNA sequence of a gene, what can we do with it? The first thing to remember is that knowledge of the DNA sequence does not automatically mean that the protein product is known; we might not know exactly where the gene starts and stops. The other piece of information we need is the 'reading frame'. This gives us knowledge of where a gene starts and which sections of DNA are used and which are not. As we have discussed, one of the many surprises to come up for gene hunters was the revelation that not all DNA in a gene was useful, or used in the production of the final protein. For years it was assumed, quite naturally, that since DNA was so important then if it was in a genome, plant or animal, it must have some purpose by coding for something. Now we know better; some DNA is interspersed within a gene and codes for absolutely nothing. This is often referred to as 'junk DNA', but it should not be inferred from this name that it has no use. In fact these apparently unused sequences may indeed be vital for evolution, allowing changes to the DNA without detriment to vital genes, and they may also help in controlling gene expression. In the more precise terminology of molecular biology, these sections of 'junk DNA' are called introns; the bits of DNA which are transcribed and translated into proteins are called exons.

The extraordinary thing about the intermixing of introns and exons is that it can make it very difficult to determine the end product of a gene, the protein that either gives us our health or causes us problems. The use of powerful computer programs in recent years has made it much easier to determine the reading frame, but it is still not an easy job.

At one time, when trying to find a gene, it was normal to track back from a known product, normal or abnormal, and finally determine the gene involved. Then in the 1980s technical developments meant that it became possible, for the very first time, to track down a gene without knowing what it coded for. This was the way in which two of the commonest genetic scourges of northern Europe – Duchenne muscular dystrophy and cystic fibrosis – were cornered and the first step towards treatment made. Once the genes had been identified for these conditions, work could start on finding methods of diagnosis and treatment. And this was even before the gene product had been defined! We knew the genes and the mutations that caused them to fail before we knew anything else about what they produced or how they worked.

In modern medicine diagnosis is usually the first step in treatment of a disease, whether genetic or infectious, because unless we know what we are dealing with, therapies tend to have all the appearances of witchcraft. This is even more important with genetic disease; trying to treat a genetic disease by looking at the symptoms can do little more than alleviate an acute episode of a condition. At least when this approach is taken towards infectious disease it can give the body time to sort itself out. In the case of fundamental metabolic problems associated with a defective gene, this is not usually possible. It is clear that until gene therapy is available for genetic diseases the best we can manage is early diagnosis, which can give an important advantage to the management and treatment of these conditions.

Using techniques of molecular biology the gene itself can be interrogated to determine the nature of the defect. The very large gene involved in cystic fibrosis has been shown to have several hundred different mutations associated with it, each of which results in a defective gene product. As we have seen, though, just having one defective gene is not necessarily a problem. People with a single defective gene are perfectly normal themselves, but are carriers and therefore have the potential to pass the defective gene on to their offspring. Only if both partners are carriers will there be the danger that one of their children might inherit both dud genes and be affected. This indicates that it is important not only that early diagnosis is made, but also that carriers are detected, allowing risks to be assessed and potentially affected pregnancies to be monitored from the outset. Interestingly the diagnostic tests available today are so sensitive that a simple mouthwash with 20ml of sterile saline will remove enough cells from the inside of your mouth for the direct test on the DNA to be made using a form of the polymerase chain reaction.

Huntington's chorea and the risks of knowing the future

The production of a test for these conditions is a major step forward, but some disorders for which tests have been developed do not have such a happy outcome. Consider the biblical quote from Psalm 39:

> Lord, let me know my end
> and the number of my days;
> tell me how short my life must be.

This was exquisitely adapted to music by Brahms for his *German Requiem*. But if you could ask that question and somebody could genuinely give you the answer, would you really want to know?

For some people this question can not only be asked, but also, at least to some extent, answered. People who have a family history of Huntington's chorea are in exactly this position. It has become possible in recent years to test for this condition, but unlike cystic fibrosis, the disease is a dominant disorder. It only takes one copy of the gene to be defective and the person will be affected. Huntington's chorea is particularly insidious because onset is relatively late compared with some conditions, middle age being quite a common time for symptoms to start. This means that children may have seen the mental and physical decline of their parents, ultimately ending in their premature death, in the full knowledge that they, too, might well be affected and this might also be their fate.

Being dominant, the probability of inheriting the defective gene, and therefore the disease, is 50/50. Under these circumstances there is an all too understandable reluctance for some people to be tested. After all, with no treatment available it might be better not to know than risk finding out with certainty the manner of your death. Others take a different view and see finding out, whether the result is either good or bad, as better than living in a state of uncertainty. It is all a matter of personal preference, something which should always be respected, by you, me and insurance companies.

Future possibilities for gene therapy

Given the difficulties associated with even thinking about how to approach altering a person's genetic make-up, the idea of gene therapy seems far fetched. It is no longer adequate to think just in terms of alleviating symptoms; at the dawn of the twenty-first century we should be looking forward to greater possibilities, either repairing a defective gene, or replacing a defective gene with a fully functional one. Again, it is not necessary to know precisely what the gene does, only what it is, where it can be found and whether it is functioning correctly.

Because of the prevalence of cystic fibrosis and the quantity of data available about this condition, it has been the subject of considerable research aimed at gene therapy. It is not the only disorder under investigation for this form of treatment, but it will serve as a general description of some of the approaches being used. The first point to understand is that there are potentially two different types of gene therapy: somatic gene therapy and germ line therapy.

Somatic therapy is where the genetic change is made purely to the constitution of an individual, their cells and tissues, but most importantly only in that individual. Germ line therapy, on the other hand, changes the genetic make-up of the cells which will form either the eggs or the sperm. As you will realize, following germ line therapy, manipulation of germ cells, all future generations will inherit the altered gene.

As soon as it became possible to consider the idea of gene therapy, it was realized that it might be setting up a time bomb in the human gene pool to undertake germ line therapy. In fact, germ line therapy was specifically banned in many countries for this reason. Unfortunately, however, nature is an intricate and determined partner and it has been reported recently that in mice that have been used to test somatic cell gene therapy techniques, some seem to have ended up with germ line alterations. This means that great care will have to be taken in genetic research; while politicians can legislate for what humans do, they cannot control genes with such ease.

Some disorders have an extraordinarily high carrier frequency; the question then has to be asked what is the value of this huge mutational load of abnormal genes? If they have a use then to eradicate the abnormal genes might just be storing up problems for future generations. For example, approximately one in 20 people in northern Europe carries the defective gene responsible for cystic fibrosis. The rate of new mutations in this condition is relatively low, so we can surmise that this load must be maintained in a population for a reason. Contrast this with Duchenne muscular dystrophy. Here we find that between 30% and 40% of cases of muscular dystrophy are due to new mutations. So what is this telling us? One gene is highly susceptible to mutation, while another, being more robust, does not mutate so rapidly but seems to be maintained with a high level of abnormality none the less. It has been suggested that one explanation for this is that carriers of cystic fibrosis, not affected people but carriers of a single defective gene, have some resistance to infection by another scourge, tuberculosis.

It is not possible to say, at present, whether or not this is true, but since we cannot be certain we would be very unwise to upset the equilibrium of gene balance by using germ line therapy. On the other hand, the possibility of treatment for affected individuals can only be applauded.

Gene replacement therapy

So far, the best line of research seems to be not to repair a defective gene, but to replace it entirely. The repair of a defective component is a far more complicated procedure than replacement. Nevertheless, it is by no means easy to replace a gene. In order to accomplish this it is necessary to introduce the gene and all its control sequences into the correct cells. This simple statement belies the extraordinarily complicated means and methods that have to be resorted to in order to achieve this goal.

A first step is to clone the gene to produce multiple copies by inserting the gene into a microorganism and growing and multiplying it. This is important to ensure that the correct gene is available in large enough amounts to allow handling and manipulation. Once cloned, it is necessary to find a carrier that can be used to insert the gene into the correct cell. Gene therapy research usually uses either a virus or an artificial carrier for this purpose. The artificial carriers are in the form of small spheres or vesicles, made up of pieces of membrane. These are able to fuse with the membrane of a living cell and discharge the DNA into the cell. This method is surprisingly effective considering the apparently haphazard nature of the process.

When a virus is used as the carrier it is normally a virus called Adenovirus that is the vector of choice. This virus normally infects the respiratory tract of humans. It may seem strange to use an infectious disease-causing virus in such research, but the logic is that if a virus causes infections by injecting their DNA into a cell, then adding the replacement of the defective gene to the virus will ensure that it, too, is injected into the cell. When the virus then goes on to replicate itself, the artificially produced gene for the normal cystic fibrosis product will also be reproduced. If this gene is also incorporated into the host DNA then the grail is achieved and normal expression of the gene product will result. In clinical terms the patient would see an alleviation of symptoms and, even better, it might also result in a cure.

The alternative method of introducing a replacement gene is to use artificial carriers called liposomes. These simple membrane balloons contain the gene and all its accoutrements to produce the gene product. When the balloon fuses with the cell membrane the contents are discharged into the cell. Exactly how the DNA is transferred from the liposome to the nucleus and activated is rather obscure, but experiments have shown that it can work. This is where the liposome system is at a disadvantage compared to the viral system of introduction. With the viral system, all the necessary information is held in the genome of the virus to incorporate the DNA into the cell and start its transcription.

There is no doubt that some body tissues are currently easier to deal with than others. Most genetic diseases are confined to a particular tissue type or organ of the body. For example, cystic fibrosis affects highly secretory systems like lungs and guts and muscular dystrophy affects muscles. It may not always be the case that genetic diseases involving single tissue types are the only ones with the potential for treatment, but it is definitely easier at the moment to treat tissues which can be easily removed from the body and then replaced. This may sound strange, but if you can remove the tissue it can be manipulated in the laboratory and then replaced back into the body.

Blood and bone marrow are just such tissues. There is a very rare genetic disorder, called severe combined immunodeficiency syndrome or SCID, which has received considerable attention in recent years. Even though the disorder is referred to as a 'syndrome' and the effects are widespread in the body, about half the cases can be traced to a single enzyme deficiency, the lack of an enzyme called adenosine deaminase. Without this enzyme the body cannot reliably defend itself against invading microbes, hence the immunodeficiency part of the disease name. SCID is known to affect only about 40 children each year in the entire world. However, bearing in mind that where conditions do not allow for early diagnosis the child dies very quickly, we can assume that this figure is very much smaller than the true figure.

By virtue of the nature of the disease, SCID is a prime target for gene therapy research and this started in 1990 in the USA and 1993 in the UK. White blood cells were removed from a patient, a corrected gene inserted and the cells replaced. Unfortunately SCID is proving, like all genetic disorders, to be a stubborn adversary and there are still many problems to be resolved before we can say that gene therapy is an unqualified success.

Sickle-cell anaemia

Having talked of replacing a defective gene with a good copy, what about a little lateral thinking? What if a person had several genes for a single product, but only one was ever active; would it be possible to switch off a defective version and switch on a working one? This is the approach that has been applied to sickle-cell anaemia and although for a long time it has been no more than an interesting theory, recent research has begun to make active investigation worth while. Let us look at sickle-cell anaemia in more detail – it has several features which make it a unique genetic disease.

Ever since the American army instituted routine testing of black recruits for sickle-cell disease in the 1940s we have been developing a better idea of the prevalence and nature of this condition. Make no mistake, it is a very large problem, although the cause is small, in fact as small as it is possible to be in genetics. A single base difference in the DNA sequence of one of the genes which codes for a haemoglobin molecule results in a single difference in the amino acid sequence. Result? The protein, the globin molecule, no longer behaves correctly. The abnormal haemoglobin in the red blood cells passes its oxygen from the blood to the tissues with the result that the haemoglobin ends up as a sort of insoluble polymer. When this happens the electric charge carried by the abnormal haemoglobin causes the red blood cells to collapse into a sickle shape – hence the name of the condition. This change in the cells is visible under the microscope and is what inspired James Herrick, a nineteenth-century medic working in Chicago, to come up with the description of the disease which is still used today.

If sickle-cell disease only resulted in a change in the shape of the red blood cells, then we would probably regard it in the same category as the ability or lack of ability to roll your tongue long-wise, but nothing more. Unfortunately, however, when the red blood cells 'sickle' they clog up the narrowest blood vessels, the capillaries, which deprives the surrounding tissues of oxygen. This oxygen stress eventually leads to organ damage and, if untreated, to organ failure or stroke. Either way, the result is death.

World-wide there about 100,000 children born with sickle-cell disease every year. In the UK there are between 5,000 and 6,000 patients, while in the USA there are approximately 65,000 sufferers. Most of those affected are African in ethnic origin. As in cystic fibrosis, this very high carrier rate has been thought to be associated with resistance to an infectious disease, in this case malaria.

Sickle-cell disease played a central role in the early days of genetic research when American biochemist Linus Pauling demonstrated in 1949 that it was possible, using electrophoresis, to separate the haemoglobin of affected individuals, normal individuals and those that were carriers of the sickle cell trait. This work was crucial in demonstrating the relationship between a defective gene and its protein product. Pauling was without doubt an exceptional person. Not only is he one of the few people to win a Nobel Prize, but he was the first person to win two unshared Nobel Prizes, one for chemistry in 1954 and one for peace in 1962.

Soon it became apparent from the mode of inheritance that sickle-cell anaemia was a recessive disorder. Carriers exhibited both the normal and sickle-cell haemoglobin, but were ostensibly normal until they were put under conditions of abnormal oxygen load, like high altitude. For the affected individuals, with two defective haemoglobin genes, the condition is much worse and in the past death usually occurred before their second birthday, either from organ failure or stroke as the sickled cells clogged small capillaries and stopped oxygen transport in vital organs.

Since the introduction in 1978 of a DNA test for sickle-cell anaemia an interesting observation has been made which has great potential for the treatment of this disorder. Some people who had two copies of the defective gene and therefore should have been affected were found to be, to all intents and purposes, normal. So what was the reason for this? The observation invoked curiosity because it was thought that if we understood this process, we might have a possible route for treatment. The story of how this strange 'affected and yet not affected' condition occurs goes back to the way we develop in the womb.

We do not simply have one type of haemoglobin. As the foetus develops in the womb the need to exchange oxygen with the mother across the placenta alters. To keep the process working correctly the affinity of the maternal and foetal haemoglobins for oxygen must be balanced and different between mother and child. Very early on the developing child produces embryonic haemoglobin; later on this is switched off and foetal haemoglobin switched on. The change to adult haemoglobin, the haemoglobin which we normally all use and which has the potential to induce sickling when defective, takes place around the time of birth, at which point if the child has the disease the first symptoms will appear. So what about those individuals who should be crippled by the disease but are not? It seems that these people never make the switch from foetal haemoglobin to the adult form and so are unaffected.

This observation offers one of the best hopes for the future treatment of sickle-cell anaemia. If affected individuals could be induced to turn off production of adult haemoglobin and turn on production of foetal haemoglobin, the symptoms and virtually all of the associated problems could be alleviated. This is a slightly different approach to gene therapy from that used for disorders such as cystic fibrosis and serves to remind us that just knowing a sequence is not necessarily enough to enable us to approach a problem in genetics effectively. Very often the aetiology of a disease, what causes it to happen, is just as important as the underlying mechanism. This has been the approach for many years in the fight against a disease at the forefront of much medical research – cancer. The problem with cancer is that by the time it is diagnosed the original causal agent may be difficult to identify and degenerative changes mask the lesion which resulted in the condition in the first place. This is one of the reasons that it took so long to recognize the link between smoking and lung cancer.

Cancer and the immune system

It should always be remembered that cancer has been with us forever, it is not modern, it is just that as the severity of infectious diseases in western society has declined, so the baseline causes of death have also changed. But do not be taken in by the belief that because we live longer so cancer becomes a greater problem. First, we can be sure that in the past the diagnosis of disease was a somewhat amateurish activity, especially when applied to the poor, resulting in generic terms for a wide range of disorders. These included 'consumption' and, perhaps more surprisingly, leprosy.

There are two major points here: the first is our mistaken ideas about longevity and disease, and the second is our mistaken ideas about the purpose of our immune system. To avoid unnecessary complication, let us take the second point first.

Simply by logical reasoning we can show that our immune system only attacks infectious organisms incidentally; its primary purpose is quite different. We know that spontaneous mutations in cells occur at a rate of between one in a million and one in 1,000 million per cell division. Now although we know that radiation and some chemicals can increase this rate, these figures can be seen as a sort of baseline rate, representing misincorporation of bases in the DNA as it is replicated in preparation for cell division. This may be thought to be a surprisingly low rate, but consider the following. There are about 1,000 million cells in a gram of tissue. Of course, not all of these are dividing, so some will never show mutations. Some tissues, however, do have a very large turnover of cells indeed. One such is bone marrow. This tissue produces blood cells, and in a human weighing 70kg, because of the limited lifespan of blood cells, the marrow has to produce something like a million million cells every day. Suddenly the rate of one in 1,000 million mutations per cell division seems not so low after all.

Once we get to grips with these huge numbers we start to see a puzzle. Under these circumstances we would expect to see far more mutations than we actually do. One of the manifestations of these mutations are tumours, so why are we not universally affected by tumours? This is where the protection provided by our immune system comes in. Our immune system originally evolved to seek out mutant cells and destroy them, protecting the body against the development of tumours. This means that the protection against infections is almost irrelevant; it was at best a secondary purpose.

Without a system for removing mutated cells a large and relatively long-lived organism such as a human would not be able to survive. If you consider what happens to individuals who have a compromised immunological system, such as in AIDS, you will soon realize that tumours are one of the burdens of this condition and would be for the rest of us, but for our immune system.

So what of the other point, that we are mistaken in our perceptions of longevity? It is sometimes said that during the middle ages the mean life expectancy was only around 30 years. Unfortunately, in the next breath an incorrect conclusion is drawn, and an old myth is perpetuated. This is that during the middle ages people did not live beyond their 30s. This is not the same thing at all!

In the UK, mean life expectancy in 1948 for males was 66.4 years and for females 71.2 years. At the turn of the twentieth century mean life expectancy is about 74.2 years for a man and 79.4 years for a woman. But next time you visit a churchyard, have a look at the gravestones. There were a lot of very old people in past centuries as well. The reason for this misunderstanding is that childhood is a very dangerous time and always has been, although in past centuries it was even more so than it is now. Death rates during childhood have been reduced considerably in recent times in developed countries. In 1948, infant mortality, that is the number of children dying between the ages of one and four years, in the UK was 5,019. By 1994 this was reduced to 796. To put it another way, take a relatively straightforward theoretical example. If life expectancy is, say, 30 years and mean age at death includes all the newborns, only 50% of which survive, then if you make it past childhood you could reasonably expect to live until you were 60 years old. This is not the same as saying average life expectancy is 30 years so people only lived into their 30s, and demonstrates the pitfalls of jumping to conclusions via a simple statistical approach to a problem.

The new science of pharmacogenetics

One of the most exciting new developments in genetics is what is called pharmacogenetics, or sometimes pharmacogenomics. Pharmacogenetics may be one of the first health-related areas of DNA research that most people will come across, and it will almost certainly revolutionize standard ideas about drugs and disease. Pharmacogenetics promises to impact on the way in which diagnosis of diseases of all sorts, including infectious diseases, is carried out.

Broadly speaking, pharmacogenetics is the study of the interactions between genotypes and drugs. At first sight this may not seem to be a particularly fertile area of investigation, but it is prompted by several observations. The first is that we are all different and the second is that up to a third of people do not respond to prescription medicines. Also there is the realization that in the USA, of 3 billion prescriptions a year there are reports of as many as 2 million adverse reactions and 100,000 deaths. This puts death due to adverse reactions as the fourth largest cause of death in the USA.

Eventually it is envisaged that diagnosis of a disease, be it genetic or infectious, will take into account the genome of the individual. This seems quite reasonable since we all have subtle variations in our metabolism that may or may not render a drug suitable for use. Until now we have been in the hands of a medical approach that advocates presciption of the pharmaceutical followed by a period of waiting to see if it works. For some it will, for some it will not, but for an unfortunate minority of patients the results may be worse than the problem that drove them to the doctor in the first place.

At the moment the technology of pharmacogenetics is still in its infancy, but is developing with great speed. The first intimations of genetics affecting the outcome of treatment came during the 1950s when it was noticed that some patients simply stopped breathing after administration of a pre-med relaxant just prior to anaesthesia. These people had a mutation that did not allow them to clear the pharmaceutical from their system and they stopped breathing. Since then it has taken a long time to clarify the association between genome and drug therapy, but there is a light at the end of the tunnel with the discovery of minute but extremely informative changes in DNA called single nucleotide polymorphisms (SNPs), pronounced 'snips'.

These very small changes are tested by using an array of probes bound to a single treated surface, or 'chip'. By having a large number of probes attached separately to a very small area in a determined and known manner, the whole test can be used as a 'dip-stick' for specific genetic variation. When a DNA sample from an individual is used, the complementary DNA sequence on the chip binds it. This can then be analysed to determine the degree of variation present in that patient. By correlating the SNP variation with knowledge of drug susceptibility it is then possible to prescribe the drug most likely to work for that person. It is already possible to use a chip to detect several variants of the gene for a substance called cytochrome P450, which has an important role in the process of energy production in the cell. Cytochrome P450 is associated with the metabolism of about 20% of drugs and some variants tend not to be so good at it. If the patient is a poor metabolizer then large or even normal doses of a drug may become toxic by slow accumulation. Ethnic variations in cytochrome P450 mean that for certain drugs about 10% of white people are poor metabolizers, compared with less than 1% of Asian people.

There are already many known cases of drug action being modified by a genetic component in the recipient. As early as the Second World War it was noticed that some soldiers given primaquine to ward off malaria developed haemolytic anaemia. This was related to a deficiency in a specific enzyme called glucose-6-phosphate dehydrogenase, which is a problem generally associated with people of African descent. Even now primaquine comes with a warning to this effect.

Knowledge of these connections and developments in pharmacogenetic research will undoubtedly make the use of drugs both safer and more effective in the future.

The development of DNA vaccines

DNA has become an important tool in maintaining our health in quite another way – the use of vaccines. Today, at the turn of the millennium, naked DNA vaccines are only just coming towards clinical trials, but the story of their development started back in 1993. In the first experiments mice were injected with DNA known to encode the antigenic protein of influenza virus, the organism that causes 'flu. Although the exact mechanism of response was not known, it was found that the mice developed an immune response to the virus. Later, in 1998, mice were inoculated with gold beads coated with RNA, rather than DNA, that had been made artificially and coded for a protein of the tick-borne disease viral encephalitis. The mice seemed to mistake the RNA for messenger RNA normally made by the virus during infection. All the mice produced antibodies and survived infection with encephalitis. It is thought that this method of approach might well yield results in diseases that currently do not have an effective vaccine, such as tuberculosis, malaria and chlamydial disease. These three diseases alone constitute a serious threat to health in many countries of the world. Chlamydia, for example, which is currently thought to infect 400 million people world-wide, is a bacteria that causes a range of problems from genito-urinary infections to blindness.

DNA vaccines are also being considered as a potentially useful and effective treatment in many diseases, including tumours. For example, results are just becoming available for a DNA vaccine against hepatitis B. The mechanism used in this early trial was to coat microscopic gold beads with a coating of viral DNA and inject these into eleven volunteers. All of the volunteers produced sufficient antibodies to give very good protection against infection. This technique is another demonstration of the surprises that DNA has in store for us. But, as with so much technology, the trial has to be approached with great caution. If the naked DNA were to be inadvertently incorporated into a person's genome it might disrupt the function of genes vital to life. In this respect it is thought that the RNA vaccines, as used on mice, might be safer, as incorporation could not occur.

While early workers thought that the study of genetics would be confined to research into genetic diseases, it has become progressively more obvious that genetics has the potential to touch all areas of disease and disease treatment. This extends from obvious genetic lesions, such as cystic fibrosis and haemophilia, to the less obvious such as cancer. With the increasing understanding of the very individual response people have to the biochemical intervention of disease, both genetic and infectious, genetics are becoming more important in deciding not only which drug to use in a particular case, but also the dosage. But the value of DNA analysis does not stop there, because it also has the potential to help in control of infectious diseases by the production of new vaccines. This should prove useful to add to the panoply of control methods against long-established diseases. With the emergence of apparently new and novel diseases this century, such as AIDS and drug-resistant TB, we can never be sure of what is around the next biological corner, so all the additional help we can get from our knowledge of DNA will be welcome.

SIX

Sexing the parrot

The determination of sex using DNA analysis

Sex is inextricably linked with reproduction but the reverse is not true; reproduction is not inextricably linked with sex. While we think of sex, in the sense of gender, as being either male or female, it is interesting to note that this is probably the exception rather than the rule in most species. Sex is not confined to animals either; some plants have different sexes as well. For the early geneticists it seemed reasonable to assume that DNA in the form of chromosomes was the major determining factor as to whether an organism was male or female. As we will see, this notion was rather too simple; life had far more tricks up its sleeve than that.

In some species of animal, sex determination can be a rather hit or miss process. The American alligator, for example, relies on incubation temperature to determine the sex of the young. Below 30°C the eggs all develop into females, while above 33°C all males are produced. In the gap between these two temperatures the eggs can develop into either male or female. This is why I describe it as a bit hit or miss. Perhaps surprisingly, the ratio of the sexes is still about 1:1 in these alligators. In fact, temperature-dependent sex determination is quite common in reptiles and they still manage to keep a balance between the sexes. Although sex reversal, changing from one sex to another, is also not unknown, the fact that a balance between the sexes tends to be maintained makes us think that there is probably a genetic component to this sex-determining mechanism as well.

Other species are just as diverse in the way they determine their sex; in fact some organisms do not have any distinguishable sexes at all. These species tend to be invertebrates, where everything from hermaphrodites to completely asexual groups are found.

Sex in fish is, as for many animal groups, largely unexplored territory. Of the 22,000 species of fish known, only about 3% have had their chromosomes looked at. Sex chromosomes in fish can be incredibly complicated, involving many different chromosomes, but overall fish have systems as varied as any group. For example, they have hermaphrodite species, bisexuality, species in which males have two different sex chromosomes, species in which females have two different sex chromosomes, and species that have environmentally determined sex. There are even unisexual species, in which females only produce female offspring.

By contrast, amphibians are relatively straightforward in the manner by which they determine their sex. Most reptiles are also rather more dependent on DNA for their sex determination, but like crocodiles and alligators there are various species that depend on temperature to determine gender. Closely related species may be either temperature dependent or DNA dependent in sex determination, implying that it is possible to switch from one mechanism to another over evolutionary time.

When it comes to mammals the story is, thankfully, quite different. Your sex and mine is determined entirely by DNA. Admittedly this process can sometimes go wrong, but generally speaking it seems to work quite well. From a human point of view it may seem an odd thing to have to resort to DNA structure and function to identify the sex of an organism, least of all a human being. And as for sexing a parrot, the question that immediately comes to mind is 'why bother?'

The problem with parrots

In the case of parrots, and indeed many other birds, there is a very practical reason for using DNA analysis to determine the sex of an individual. If you want to breed parrots, it is important to know whether you have two of the same sex, or one of each. In some species of parrot, although the birds themselves can tell male from female, the sexes are identical in appearance to us. This inability to determine the sex of a bird is of immense financial importance to the breeders.

Having been privileged to see wild flocks of cockatoos in Australia, I would rather not have these magnificent parrots in captivity. Such intelligent and sociable birds, even when they behave like vandals breaking up trees, are wonderful to observe in the wild and it seems cruel to confine them to a cage. Parrots are generally regarded as amongst the most intelligent of birds. It seems reasonable to assume that just as solitary confinement for a human prisoner is mentally disturbing, it is probably no less so for an intelligent and sociable bird like a parrot.

The world-wide trade in parrots is a multimillion-pound business. Some species of parrot can cost several thousand pounds to purchase, so it constitutes a big investment to the breeder when a deal, legal or illegal, is struck. Although it is to be hoped that some of the very rare species which are being captive bred will not be just for the domestic market but for release back into the wild, unfortunately we know the sorts of money involved in the buying and selling of birds partly because of the enormous sums paid for illegally exported species. For example, a breeding pair of Major Mitchell cockatoos (*Cacatua leadbeateri*) from Australia, which are charming pink birds, can fetch up to $15,000 a pair, red-tailed black cockatoos (*Calyptorhyncus magnificus*) are priced at $20,000 and, most staggering of all, a pair of glossy black cockatoos (*Calyptorhyncus lathami*) might change hands for $50,000. Comparable prices are also found for rare birds of prey, though in these cases they are usually bought for hunting rather than just for show.

If you have ever had the pleasure of seeing a flock of parrots, and they are not by nature solitary birds, you will understand my enthusiasm for this experience. The loss of any species to extinction is always an incalculable shame, but it is difficult to express the magnitude of loss if even a single species of such intelligent birds as parrots should disappear. This has happened all too frequently in the past; perhaps fortunately some of the events and times have been well recorded, and it is possible we might learn from our losses. Just a few of the recent extinctions are the yellow-headed macaw (*Ara gossei*), which became extinct in 1765, and the green and yellow macaw (*Ara erythrocephala*) in 1842, both from Jamaica, the Dominican macaw (*Ara atwoodi*) in 1800 and the Cuban macaw (*Ara tricolor*) in 1864. These are just a small sample of the parrots, now gone forever, hunted, collected or just harried to extinction.

If parrots are to be bred in captivity, either to discourage their collection from the wild or to provide stocks for release into the wild, it is necessary to be able to sex them accurately. This becomes even more important if there is only a very small group of parrots, perhaps comprising two or three individuals, available. In some species, of course, distinguishing between the sexes is relatively easy; some males and females are strikingly different in appearance. For example, in the eclectus parrot (*Eclectus roratus*) from Australia, the two sexes do not even look as though they belong to the same species. For many parrot species, however, the sexes are identical in appearance.

This problem also faces zoos and breeders attempting to breed other species of birds as well, not just parrots. Many different bird groups are morphologically the same, with some of the most endangered being the most difficult to sex, such as birds of prey. It is often said that the sexes of birds of prey, the raptors, are different sizes, with the females being the larger of the two. This is, however, a very questionable rule of thumb. Once bred in captivity, or with a very small number of individuals in the wild, the sexual dimorphism of size can become obscured, and even when it is not, sexing individual birds on the basis of a statistical observation would be extremely inaccurate. One of the two sexes may tend to be larger, but the size range between the two sexes will inevitably overlap, making it impossible to be sure of the sex of any individual on size alone.

Of course, some might say that the easiest way to sort out this problem is to just ask the bird. This is much more easily done in the wild than in captivity. Parrots themselves usually distinguish between male and female through subtle differences in behaviour.

Males and females will usually behave differently, even if they look identical, especially when it comes to the behaviour associated with searching out nest sites and mating. But birds in captivity do not normally have the space for this sort of expression. Their behaviour is also altered when they are removed from a flock and put in solitary confinement, making it almost impossible to determine the sex from the activity of the individual.

Date palms and random amplification

There are many times when determining the gender of an organism becomes important, especially with financially important species. The need, or at least the desire for commercial reasons, to determine the sex of an organism is not simply confined to animals. Take, for example, date palms. These have been cultivated by humans for thousands of years throughout the Middle East; they even appear in ancient Egyptian books of spells. In fact the date palm has been cultivated for so long that it can no longer be thought of as a truly wild species.

The commercial date palm is a member of the genus *Phoenix*, which comprises approximately 17 species ranging from the Canary Islands, through Africa, India and Sri Lanka, the Malay Peninsula and South East Asia up to Hong Kong. All the species are difficult to identify accurately because of their close similarity in appearance, besides which it would seem that many of the cultivated plants are crosses between different species from the same genus.

Now, all date palm species have separate male and female plants; they are what is known as dioecious. This means that when date palms are grown as a crop, female plants are wanted more than male because it is the female plant which produces the fruit, but of course they must be pollinated by pollen, which can only come from a male plant. Because the male flowers produce such huge amounts of pollen, it takes only a single male tree to pollinate large numbers of female flowers. Ideally, therefore, growers want to have mostly female plants with just a few male plants for pollination.

A natural population of date palms, however, would, of course, produce 50% male and 50% female seedlings. A crop with these proportions would be unproductive. Clearly, the grower needs to be able to determine the sex of a seedling and to weed out the males. Unfortunately, in the past it was only possible to determine the sex of a date palm when it flowered. By the time the grower knew which date palms were male and which ones female he had spent more than ten years nurturing the seedlings.

How much better, then, to be able to determine the sex of the seedlings before they are planted out, so that the plantation owner can choose how many female seedlings to plant, how many male, and how they should be distributed around a plantation. In a fortuitous observation it has recently become possible to sex date palm seedlings using that most powerful of molecular biology techniques, the polymerase chain reaction (PCR). With a set of primers which bind to specific sites in the genome it becomes possible to amplify short sections of DNA which can then be separated on a gel and visualized by fluorescing them with ultraviolet light.

Now, because date palms have not been subjected to extreme breeding programmes, which is different from saying that they have been cultivated domestically in plantations for food, there are many differences between the two sexes that can be detected in the sizes of these amplified fragments when they are separated on an electrophoresis gel under the influence of an electric current. Indeed, there are so many differences between the sexes that a technique called random amplification of DNA (RAPD) can be used.

Although based on the polymerase chain reaction and described as random amplification, this is something of a misnomer. The sequence of the primer may originally have been random, with someone simply making up a primer sequence to see what would happen. However, once a primer sequence is found, every time it is used the results are both predictable and repeatable when used on the same species. Where a random primer sequence is used and, with luck, a difference in the products between the sexes is detected, we have the basis of a PCR test for determining sex. This is true whether it is a plant or animal that is being tested. In the case of date palms this technique allows for a very small piece of the seedling to be used to determine the sex, so small in fact that it causes no damage to the plant. This gives the growers a great advantage in cultivation.

Applications to humans

With other species the reasons for needing to know the sex of an individual can be quite different from those in the date palm or parrot. Take, for example, humans. Currently there are many very good reasons why it is not only useful but also essential to have a reliable method of determining a person's sex, especially prenatally or soon after birth. Of course there are many difficult decisions concerning the use of information to do with inherited genetic diseases, whether they are sex-linked or not. In the event, such stressful situations may involve genetic counselling, but in reality the final choice can only be made by the parents concerned. On this basis we have to accept that sometimes difficult decisions have to be made that we do not necessarily agree with. Given the information that a child will be affected by a sex-linked genetic disease, what we would ideally like to be able to do is correct the deficiencies and genetic lesions, especially those which result in a life-threatening condition.

At the moment it is not possible to correct a genetic problem, although alleviating the symptoms might be possible in some conditions, either by changes in behaviour or, in the future, by gene therapy. Unfortunately even if it does seem an unpleasant option, sometimes the only answer is to terminate an affected foetus as no treatment is available and early death certain. We should not be hard on those who have to make these choices; a family that has had one child incapable of independent sustained life will be understandably reluctant to produce another.

A lethal sex-linked disorder can be so devastating for a family that after one affected child a family may decide not to have any more children, or alternatively, may decide to terminate pregnancies that might be affected. The important phrase here is 'might be'. This is not a decision that is ever taken lightly and the weighing up of odds of having another affected child is not only difficult, but also a subjective one. It is easy to calculate a figure for the probability of inheriting a sex-linked disorder, but not so easy to assess the problems the family will have to face.

Without a specific diagnostic test the only way of shortening the odds for a family would be to identify the sex of the foetus. In the case of something like haemophilia, knowing that the foetus is male makes the calculation of the chances of inheriting the disease simple – it will be 50:50. In this case to avoid any chance of producing a child with haemophilia the only option would be to determine the sex of the foetus in the womb and to terminate the pregnancy if it is found to be male. The situation is complicated by the knowledge that prenatal sampling techniques in themselves carry a residual risk of damage to the foetus.

Inheritance of sex

The calculation of that 50:50 chance of a male being affected once it is known that the mother is a carrier is a reflection of the way in which our sex is determined. Although some disorders, like Duchenne muscular dystrophy, have such high mutation rates that they can appear spontaneously in a family with no warning, most sex-linked disorders are more predictable.

Sex linked conditions are inherited on the X chromosome. As we discussed earlier, males have one X and one Y chromosome, females have two X chromosomes. This is the sex-determining mechanism in humans. Any defective gene on an X chromosome varies in its outcome depending on two things: whether it originates from the mother or the father, and whether it is passed to a son or a daughter.

At this point perhaps it would be useful to look at how sex is inherited in humans and other mammals. It is slightly different in other types of organism, but the attempted outcome is always a balance in the numbers of males and females.

Most humans have the same number of chromosomes in every cell – 46. These comprise 22 matching pairs with two additional chromosomes left over. In women these two remnant chromosomes are a matching pair, while in men there is one large and one small chromosome. The matching pair are called X chromosomes and the non-matching pair are an X and Y, the Y chromosome being very much smaller than the X chromosome.

While an egg from a woman will always contain an X chromosome, sperm from the father can contain either an X chromosome or a Y chromosome, but not normally both. It is generally regarded as a random process as to which of the sperms receive an X and which a Y chromosome. It is also regarded as more or less random as to which sperm does the fertilizing. It is no surprise, therefore, that the overall ratio of the sexes is 50:50. When sperm meets egg, half the time the products will end up with two X chromosomes, one from the egg and one from the sperm resulting in a girl, and half will end up with one X, from the mother, and one Y chromosome from the father, resulting in a boy.

Testing for chromosomal abnormalities

In the early days of prenatal testing in the 1970s, the standard method of testing was to take a sample from the foetus using an indirect method called amniocentesis. This process is normally carried out by removing some of the amniotic fluid from around the foetus. This fluid contains cells that have been naturally shed by the foetus and these can be cultured in the laboratory. Great care has to be taken when taking the sample that the needle does not come into contact with the growing foetus. It is also important to take only a very small volume of fluid as it is the amniotic fluid that acts as a shock absorber for the baby in the womb.

Once the sample has been taken, the cells in the amniotic fluid can be cultured so that they grow and multiply in the laboratory. Strictly speaking, the cell colonies which result from this process of cell culture are called clones as each colony starts from a single progenitor cell, but this is most definitely not the same as the generally accepted idea of clones and cloning. Cultured cells are then treated so that the chromosomes can be examined to determine exactly which sex chromosomes, X and Y, are present.

Sex determination is not the only reason for using amniocentesis, in fact it is quite a rare one. The most common use is to test for inherited conditions in which abnormal numbers of chromosomes are present, such as Down's syndrome. Unfortunately, because of the technical constraints of cell culture, amniocentesis is a very slow process and the wait for the results can be an extremely distressing time for a family at risk.

More recently, however, a new technique has been developed for prenatal testing which makes it possible to remove a little piece of material from around the edge of the growing placenta, which is foetal in origin. This process, called chorionic villous sampling, provides material in which, unlike cells from amniotic fluid, the cells are already actively dividing. This means that no culturing is necessary and a result can be produced very much sooner than is possible with samples from amniocentesis.

Whether the sample has been produced using amniocentesis or by chorionic villous sampling, the next stage, which is to actually look at the chromosomes and identify them all, is the same. Under the microscope, it is not only chromosomes X and Y that are individually identifiable. Every pair of chromosomes can be chemically treated so that they are distinguishable from every other pair. This identification, as one can imagine, is not easy, requiring a skilled technician to sit at a microscope for long periods. While this sort of analysis is routinely carried out for disorders such as Down's syndrome, it does take considerable time, money and skill. As can be appreciated, it is the time aspect of these texts which is most distressing for the parents.

Determination of sex in the newborn child

Sometimes, when there is a suspicion of a major chromosomal disorder or doubt about a child's sex, a blood sample will be taken from a newborn baby for similar analysis. This latter situation – the need to confirm a newborn baby's sex – is not as unusual as you might think. Even with sex being determined genetically, the development of those characters which tell us whether a child is male or female are the product of a very complex embryology. Sometimes this goes wrong and the outcome can be ambiguous. In these circumstances, it is necessary first of all to establish the sex of the child and then surgical reconstruction can take place. It is at this

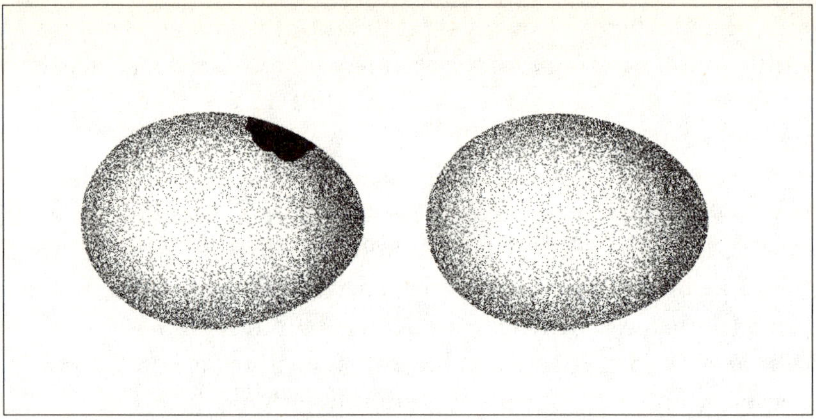

Barr bodies. Only found in female cells (left), a Barr body represents a contracted and unused X chromosome. Under the microscope it appears as a dark area on the nucleus. Male cells (right) only have one X chromosome and so do not have Barr bodies.

stage that time is so important; waiting to know the answer is both stressful and potentially embarrassing for the family. So what can be done to speed up the process?

One method that has been used for many years is quick, easy, does not require the taking of a blood sample and can be done at any age. First, a sample of cells is taken by gently scraping the inside of the cheek with a wooden tongue depressor, something like a lollipop stick. These are smeared onto a microscope slide and viewed directly to determine whether a specific feature called the Barr body is present. Barr bodies were first described by Murray Llewellyn Barr in 1949. While he was working in Ontario he discovered that there was a small spot visible on the nucleus of female cells. This never appeared in male cells and thus provided a specific diagnostic test for sex. It turned out later that the Barr body was, in fact, one of the two X chromosomes which are present in females (see the illustration above). Of course, this method of sex determination is not possible in animals such as reptiles and fishes, which do not have the straightforward XX/XY method of chromosomal sex determination found in mammals.

The amelogenin gene

Sometimes, simple methods such as looking for Barr bodies or counting chromosomes are just not feasible. For example, you may want to know the sex of a person from a blood stain left at a crime scene, or to carry out a retrospective study where the sex of a sample may have a bearing on the results but the sample is too old for traditional methods of analysis. With samples like these we can exploit another aspect of genetics which utilizes the smaller scale of DNA.

The Y chromosome is really quite extraordinary; it is not only responsible for maleness, but it is also implicated in most violent crime. After all, most, although not all, violent crime is perpetrated by men, as are virtually all rapes. Even so, the human Y chromosome seems to be almost completely inert genetically. What it does do is start a cascade of actions by all sort of disparate genes scattered across many different chromosomes throughout the genome which finally results in all those characteristics associated with maleness. This whole process seems to be instigated by a single gene that is sacrosanct from mutation. Any mutation of this starter gene would thwart the ability of a species to define two sexes.

By an extraordinary coincidence, there is also on the Y chromosome a gene which is a copy of one found on the X chromosome. The strange thing is that the gene on the Y chromosome is both non-functioning and of a different size. This gene is called amelogenin, and detection of the gene forms the basis of a sex-determining test used in forensic science. This identifies the sex of a person from a stain left at the scene of a crime.

Since with a little bit of skill it is possible to detect differences between genes, we should be able to differentiate between a male and a female version of the amelogenin gene because of their difference in length. Indeed, this is quite straightforward. The functioning amelogenin gene produces a protein associated with the development of tooth enamel. Due to a strange evolutionary process which is not yet understood, the gene, in different forms, appears on both the human X and Y chromosomes. One of the differences between them is that the Y chromosome version is shorter, having had some DNA deleted. This stops it being a functional gene. So if we measure the length of the genes we can determine the sex of the sample donor and this becomes the basis for a sex-determination test. If there are two different sizes of genes then the sample is clearly male, one gene from the X and one from the Y chromosome, and if there is only a single size of amelogenin gene present the sample comes from a female. Probably the most surprising thing about this is that although the Y chromosome version is apparently non-functional it does not appear to have undergone any further mutations or deletions, so the results are always predictable.

Like all ideas linked to clear logic and clear results, the pathway is strewn with broken ideas, and sex determination using PCR of the amelogenin gene is no exception. Anyone who has worked in clinical human genetics will know of at least one case where an individual was apparently of one sex, at least in terms of what they looked like, but genetically they were the opposite. What should we make of that? Certainly we should not junk all our theories of sex determination, which have been built up over a century. Neither should we disregard our hard-won knowledge that determination of sex is genetic. In these cases we should be looking more at the factors that can upset the normal process of early sexual development, such as hormone balance in the womb.

As discussed earlier, determining the sex of an individual might be desperately important in situations where no treatment for a sex-linked condition is available. It is impossible for anyone who has not lived through the heartache and trauma of coping with a disease such as muscular dystrophy to know what it is like. Parents who have brought one child into the world with this devastating disease are faced with many difficult decisions. Some will decide not to have any more children at all; some will weigh up the probabilities and judge the risk of another pregnancy worth taking; and some will take the option of having prenatal sex-determination tests carried out in future so that any male foetuses can be terminated. It ill becomes us to censure any of these decisions. Unless we have been in that position ourselves it is not possible to appreciate just how unkind life can seem when you watch an apparently healthy child become wheelchair bound as his muscles waste away until he reaches the point at which breathing is no longer possible.

Considering the possible outcomes, it is wrong for outsiders to judge the decisions families have been forced to make. We should, at the very least, be able to see the value of determining the sex of an individual. We might not condone the outcome, but we do have to understand it.

Genetic imprinting

A curious phenomenon has recently arisen in genetics as a method of explaining some very strange observations. This phenomenon is called imprinting. This should not be confused with the term used in animal behaviour. Behavioural imprinting, as described by Konrad Lorenz, is the process in which a newly hatched bird will regard the first relatively large moving object it sees as a member of its species. Genetic imprinting, on the other hand, is quite different and has far-reaching effects, not only on individuals but on the whole of society.

An imprinted gene or whole chromosome is marked so that the origin is recognizable, coming either from the father or from the mother. Perhaps bizarrely, the first reports of this appeared in the 1960s in a fly, but it quickly became apparent that it was not just the gene sequence that was important, but also which parent contributed it. Any early observation of imprinting was in triploids. These are individuals that have three complete sets of chromosomes instead of the normal two. This condition is invariably fatal, but the way in which death occurrs differs, depending on whether the additional set of chromosomes originates from the father or the mother. If the additional set is maternal, the placenta does not grow properly and early miscarriage results. If the additional set of chromosomes is paternal, the placenta is highly overgrown and the foetus underdeveloped.

Although imprinting has significance in disease states it also has significance for us all in our society. Why should such a phenomenon occur at all? This area still remains theoretical, but there are indicators as to why this happens and why there might be differences between boys and girls.

The story begins with the realization that sexually reproducing species generally have two complete copies of all their genes, one maternal and one paternal. Now, some genes carry information which enables them to be recognized by the cell and their parental origins worked out. Not only that, but although most genes are expressed in both copies, some genes are switched off and only those versions from a specific parent are expressed. The mechanism that causes this switching off comes from a really rather straightforward and small chemical change to the DNA. This gives the apparent, but quite erroneous, impression that some genes 'know' their parental origin.

There are some very interesting ideas regarding sexual differentiation and evolution that can be gained from imprinted genes. A very good example of this is found in a gene known as *Igf-2*. This gene produces a growth factor which influences the growth of the foetus. This gene is switched off in the foetus when the gene comes from the mother, but switched on when it comes from the father. It is suggested that this is because it is in the father's interest to promote growth of his foetus at the expense of the mother because future children are likely to have a different father. It is necessary to remember at this point that most human populations tend towards polygyny. Given that monogamy seems to be the exception rather than the norm, a father would want the most robust offspring possible to perpetuate his genes.

The next question to ask is how does imprinting affect sex determination. The answer is that in a direct way, it does not. What it does do, though, is to modify genetic effects on an individual. Early on in research into imprinting it was noticed that children with Turner's syndrome exhibited two broadly different behaviours. Turner's syndrome is a condition in which only one X chromosome is present, instead of either two X chromosomes in females, or one X and one Y chromosome in males. Recent research suggests that the occurrence of these two behaviours is connected with imprinted genes. A crucial observation was made when it was noticed that when a Turner's syndrome patient inherited her single X chromosome from her mother, her behaviour was very much more socially disruptive than normal. It was more like that of boys of the same age. Thus although the child was physically female, she could be thought of as behaviourally male. On the other hand, when the single X chromosome originated from the father, the Turner's patient exhibited behaviour which if not exactly normal, was very much closer to that expected from a girl of the same age.

It would seem that at least some of the genes responsible for the symptoms of Turner's syndrome are imprinted. It is likely that this process evolved because whereas for most other genes there are two copies, genes on sex chromosomes, remembering that the Y chromosome is virtually devoid of active genes, are only present in one copy in males and two copies in females. Since over- or under production of gene products can upset the delicate balance among many other genes and their products a means has to be found to redress the balance. This is done by silencing one of the two copies. Since there do seem to be behavioural differences between boys and girls and a father always contributes an X chromosome to a daughter, but a mother contributes an X chromosome to both sons and daughters, a mechanism seems to have arisen in which some or all of the genes on the appropriate X chromosome can be silenced, or imprinted. Precisely why such differences between boys and girls have evolved remains an unanswered question.

Returning to parrots

So now we have methods suitable for determining the sex of a human, but what about the parrots with which we started? Until recently the only way to identify the gender of some parrot species was by endoscopy, an intrusive surgical procedure which includes some considerable risk to the life of the bird. Using endoscopy it is possible to look at the internal organs of a bird and to determine whether it has ovaries or testes. The next method of finding out the sex of a bird was quite an advance and made use of the growing body of knowledge about sex-determining systems in other organisms. Extensive investigation of avian chromosomes told us that not only did birds have very large numbers of chromosomes, but they also had, just like humans, sex chromosomes that could be recognized individually.

To distinguish them from their mammalian counterparts the sex chromosomes in birds are referred to as Z and W. While, as we have seen, male mammals carry two clearly different sex chromosomes, XY, and females two similar sex chromosomes, XX, in most birds it is the other way around. Male birds are described as having ZZ and females ZW chromosomes. I say most birds because in that extraordinary bunch of birds the ratites, which includes ostriches and emus, the sex chromosomes are the same size, although it is possible to distinguish them using various staining techniques.

This research opened the way for an entirely safe, non-surgical method of sex determination in parrots. By taking a blood sample or the tip of a growing feather which, like a hair root from a human, contains living cells, cell cultures could be grown using the same techniques as described above for use in human amniocentesis tests. Thus from a few cells enough actively dividing material can be created to allow the investigation of the chromosomes directly.

In a few cases, using a blood sample or plucked feather may not be a practical proposition for various reasons. This concern led to the development of another technique, originally designed in a desperate attempt to save a species that was as near extinction as it is possible to get.

Spix's macaw and the plucked feather

The most recent method of determining sex in parrots was initially used on the last wild Spix's macaw, *Cyanopsitta spixii*. Yes, the very last specimen in the wild. Humans at it again, collecting birds for their appearance and destroying wild populations. This is particularly sad because there are, or were, about 30 in captivity, far more than the number in the wild. Remember that parrots are not only intelligent, but also gregarious. If we are lucky we will never know what it is like to be the last of our species on earth, but this wild macaw did.

Spix's macaw is a distinctive, pure blue and very beautiful bird; it is also the only species in the genus *Cyanopsitta*, so at extinction we lose not just a species but an entire genus. It is the activity of hunters collecting these birds for captivity that has resulted in this species being taken to the brink of extinction. With so few individuals left, both in captivity and the wild, it may just not be possible for the species to survive. It is often said that the bird's beauty was its own worst enemy, but I would dispute such a fatuous notion. It was human greed and stupidity at fault; nothing more could justify collectors paying the best part of £25,000 for a single bird with no consideration for the consequences.

Spix's macaw lived, never in very large numbers, in palm groves and woodland edges around the centre of Brazil. It was thought that the last wild bird was a male, but it was only when two scientists in Oxford, Richard Griffiths and Bela Tiwari, managed to sex it that this was confirmed. Using a system originally designed to sex chickens, it was possible to devise a method of sexing virtually any bird species without resorting to surgery or the collection of viable cells.

What Griffiths and Tiwari did was to collect naturally moulted feathers from the last wild bird and then extract DNA from them. It took two years to collect enough feathers for the analysis, but finally there was sufficient material. Using their knowledge from work on chickens they knew there was a highly conserved sequence on the two sex chromosomes which differed in a very minor detail between the Z and W chromosomes. By amplifying the whole sequence of DNA, using the polymerase chain reaction, it became possible to complete the analysis by digesting the product with an enzyme to give different sized fragments on a gel. It is at this point that the differences appear. Breaking the product at a specific site with the enzyme reveals the differences in sequence between the two chromosomes. The fragment from the Z chromosome is 104bp long and that from the W chromosome is 73bp. It can be seen from this that as a female has a W and Z chromosome there will be two products, one of 73bp and one of 104bp, but a male with two Z chromosomes will have a single product of 104bp. When separated on a gel under the influence of an electric current and suitably stained the result is clear: a sample of male DNA will show up as a single band, whereas that from a female will appear as two bands.

In the case of Spix's macaw the outcome of all this work is probably going to be of no avail. Science cannot answer all questions, and the resurrection of a species from such a very small number of individuals will almost certainly be doomed to failure. This is where we start to unravel the reasons for the old taboos surrounding incest. These have existed in human culture throughout history and yet until recently we had no clear idea why we adhered to them. It turns out that there are clear biological reasons why it is inadvisable for closely related individuals to breed together. The main one is a phenomenon known as inbreeding depression.

Inbreeding depression and incest

All organisms investigated so far have been found to carry a number of lethal mutations, and although carrying such a mutation, especially a recessive one, does not necessarily affect the individual, it does have implications for its progeny and thus the health of the species as a whole. If an individual with a particular mutation were to mate with another having the same mutation, one in four of the progeny would be expected to be affected with the condition, two of the progeny would be carriers and one would be normal. If we take the case of our small population of Spix's macaw we can see that the outcome might be disastrous. Since it is thought that any individual carries more than one potentially lethal mutation, the smaller a founding population is, the more likely it is that it will be unable to deal with the mutational load of the population. If only a quarter of the offspring are lethally affected, the species as a whole is giving up a quarter of its breeding potential. This is an enormous stress on a species that has to deal with the rigours of an existence in the wild.

The same considerations must be taken into account with captive-bred species, and brother to sister matings have to be avoided to stop the appearance of lethal recessive genes. The genetics of the system of inheritance is, as one might expect, more complicated than this. Sometimes many genes combine to give a single result; different mutations do not need to affect the same gene to cripple an organism, or impair the reproductive potential. So where does this leave our Spix's macaws? The answer is that it leaves them very badly placed in genetic terms, and sadly unlikely to have a long-term survival as a species. The gene pool from which we are attempting to resurrect the species is probably just too small.

Nevertheless, this does not mean that we should not try to help the species survive. For one thing there may be enough diversity in the collective genome of this bird to guarantee its survival, even though on balance it is thought unlikely. Only time will tell, but there may be both a lesson and hope from a much smaller bird called the Chatham Island black robin from New Zealand.

The case of the Chatham Island black robin

The Chatham Islands lie about 850 kilometres off the east coast of New Zealand. Back in 1980 there were only five remaining Chatham Island black robins, but by 1997 these had increased to just about 200. It was intervention by humans that helped in this recovery. It was decided to remove eggs from the robins' nests and foster them to tomtits. This encouraged the robins to produce more eggs so that three broods could be raised in a year, rather than just one. The population of robins was so incredibly inbred that using all the currently available methods of DNA profiling it was only possible to show that they were all genetically identical at the sites of DNA compared. Whether this is true throughout the genome remains to be seen; some areas of the genome need to be diverse for the health of the population. However, for the moment, one thing is certain, the Chatham Island black robins are thriving, so let us hope that Spix's macaw can pull off the same spectacular comeback.

SEVEN

Our hopes and fears for the future

The truth behind genetic engineering and biotechnology

Whenever the subject of genetic engineering is discussed, the question of ethics also arises, and quite rightly so. We should not usurp the position of God by thinking that we can control life, even if we do not believe such a being as God exists. As scientists we are still trying to find our feet when it comes to the potential problems and possibilities of genetic engineering and biotechnology. What we do know, however, is that genetic engineering is going to have an important and significant long-term impact both in the form of gene therapy for inherited diseases and in the alteration of food crops, both plant and animal.

But there is a downside to this, and the debate as to the merits or demerits of genetic engineering bring up strong emotions on both sides of the argument. Let us start at the beginning and look at this issue in fundamental terms. Nobody, not even the most crazed of archetypal scientists, wants an organism released into the environment that could cause an ecological disaster. Neither do scientists want to play at being God and create a super-race, whether human or animal.

Before the advent of biotechnology, humans had been modifying natural reproductive processes for centuries in the search for better and more productive crops and livestock. Look at cloning for example. There are, and have been for many years, forests which consist almost entirely of trees cloned from a single plant. This has been possible by the familiar low-tech gardeners' trick of taking cuttings – nothing more than that.

Early successes and failures

Until recently it was true to say that no results from genetic engineering experiments had produced any alteration to an organism, plant or animal, that would not have been possible using traditional breeding methods. This is an important point. Genetic engineering originally started as techniques that merely speeded up what could be achieved using older methods, or even what would occasionally occur in nature unaided by the human hand. In fact, early attempts at using biotechnology to modify plants just did not work. For example, there was an attempt to produce a blue-flowered rose by introducing a gene from blue petunias. This failed miserably, the rose persistently producing rose-coloured flowers. Similarly, in an attempt to produce 'easy-pulp' trees it was considered essential to disable an enzyme that was supposed to be responsible for the construction of the lignified cell wall in trees, lignin being the polymer that gives trees their woodiness. Unfortunately, this experiment was also doomed to failure. Although a wild mutant pine tree was found that did not have the enzyme, it continued to grow and produce lignin, just like all trees. It would seem that merely introducing a gene into an organism does not automatically produce the desired result.

This is not to say that genetic engineering and biotechnology should be cold-shouldered. But it is important to remember that biochemistry and genetics work together and interact in many ways to produce a complete organism. Life has a multitude of surprises up its sleeve, and I am sure that many more will need to be revealed before we have a genetically engineered blue rose.

Incidentally, it is already possible to buy a genetically engineered mauve carnation in some countries. This contains the very same blue petunia gene that failed in roses. Why it should work in one plant and not the other is a mystery, but it does hold out hope for the blue rose in the future. All that is needed is human determination to unravel the story.

Some spectacular successes have already been achieved by genetic engineering. For example, by inserting a gene from a fast-growing weed into a crop it is possible to create a plant that can be harvested in only a tiny fraction of the time it takes normally. When a gene called *LEAFY*, which originated in a weed known as common wall cress, was introduced into aspen trees they were persuaded to start flowering when they were only a few centimetres tall. Similarly, oilseed rape can also be induced to flower early by introducing an extra gene. This sometimes makes it possible to grow these crops in areas where the season is too short for normal flower and seed development.

Traditional breeding

There is sometimes considerable alarm expressed by the general public at the thought of genetic manipulation of animals on the subjective grounds of cruelty. I have little sympathy with this view: if this is so important then the first area that should be considered in terms of cruelty must surely be that of the old-fashioned animal breeders. Without some consistency in attitude and action no credibility can be given to desires to do away with a product simply because of the manner in which it was produced. There is little doubt that traditional breeding methods have created dogs congenitally susceptible to dislocated hips and cancer. Some varieties also seem to have difficulty breathing, eating and walking. In this, human breeders have removed the survival of the fittest from the equation and as far as we can tell they all originated from the same species. No modern genetic engineering has been used here, just traditional breeding. Indeed it is, as yet, impossible using genetic engineering to produce any creature quite so grotesque as has been produced by dog breeders using selection based on spurious notions of aesthetics.

What has changed with the advent of biotechnology is that while natural hybridization and mutation, both between and within species, still carries on, this has an end-point of potential. This simply means that there eventually comes a point where a species barrier is found. For example, while crosses between different species and even different genera of orchids is possible, though rare in nature due to ecological rather than genetic barriers, it can be easily appreciated that it is not normal for a cross to happen between an oak tree and a grass plant. Orchids are a particularly interesting group because there are approximately 17,000 species in 750 genera but humans have created more than 50,000 hybrids without recourse to biotechnology and genetic engineering.

Crossing the species barrier

Crosses between species and genera in nature are restricted by several things. The first of course is a matter of common sense, that is simple incompatibility. Transgenic sexual behaviour as extreme as, say, the couplings between an ape and a human are hardly likely to produce viable progeny. And although the mating of a horse and a donkey does result in living offspring, they are usually sterile. This is primarily because of the difference in chromosome complement between horses and donkeys. Horses have 64 chromosomes, donkeys have 62 and thus the cross – a mule if it is a cross between a male horse and a female donkey, or a hinny if it is between a male donkey and a female horse – always has 63 chromosomes. This usually results in sterility, although it has been known for a female mule to produce a fertile ovum and give birth to either a horse or a donkey, depending on the species that fertilized the ovum.

A second factor that separates species is rather more subtle. If two species are genetically compatible, they may still be divided by ecology. This is seen in orchids where ecological niches are just as important as genetic barriers to cross-fertilization. After all, if species never normally meet they will never interbreed, so the status of the species is not going to be compromised. But there is more to this than mere geography. Sometimes, for example, orchid species that are theoretically capable of producing interspecific hybrids and that live in the same geographical area just do not hybridize. This is usually because of the remarkable and intimate relationship between the orchids and their insect pollinators, which has developed to the point where only a single species of insect can transmit pollen from one flower to another, and the insect only visits a single species.

Given these barriers to interspecific crosses, how much more unlikely is it, then, that totally foreign genes, not even from the same genus, should appear in a completely different species? Well, yes, this is very unlikely, especially in crops that have no local wild relatives, as is the case in cotton, one of the commonest engineered crops grown in the USA. But the story is not so straightforward in crops that do have wild relatives, like all the cereals which are grown. The commonest reason given for wanting to introduce a single gene into a crop plant is to make it resistant to weedkiller. This would allow powerful herbicides to be used to eradicate the weeds without affecting the crop plant. In this case, the question to ask is whether these crops could cross with wild relatives and pass on their herbicide resistance genes. Since herbicide resistance has been bred into several crop plants using traditional breeding methods and there are several recorded cases of these genes spreading into wild relatives of the crops, it seems that the answer is yes. What this means for genetically engineered plants is that it seems reasonable that this jump between species will occasionally take place.

Keeping a lid on herbicide resistance – the problems

It is not just jumping genes that might be a cause for concern in engineered crops. Take, for example, the widely grown crop called oilseed rape. Oilseed rape has the dubious honour of being both a nuisance weed in some areas and a valuable crop in others. And now that a herbicide-resistant form has been developed, it is proving extremely difficult to confine it to the areas where it is wanted. Seeds from these plants are not only easily dispersed by animals and birds, but also accidentally by those who are harvesting the seed and transporting it to factories for processing. One farmer's herbicide-resistant crop quickly becomes another farmer's herbicide-resistant weed.

It is patently silly to assume that freely seeding crops can be grown in sufficient isolation to stop them spreading beyond their boundaries, either as complete plants or just their pollen. It has, in fact, already been shown in France that hybrids between herbicide-resistant oilseed rape and wild radish can be herbicide resistant. This was found in an experiment in which a plot of wild radish was planted next to a plot of genetically modified oilseed rape and the progeny of cross-pollination monitored. The strange hybrid seems to have gained 19 chromosomes from the rape, *Brassica napus*, and nine chromosomes from the wild radish, *Raphanus raphanistrum*. Luckily, persistence of this herbicide resistance does not seem likely because these two plants are in different genera and subsequent generations had a range of chromosome numbers from 20 to 60. This makes it unlikely that they will succeed and spread in the wild.

There are currently a number of genetically altered plant crops awaiting approval; some are produced for herbicide resistance and some for insect resistance. Insect resistance would seem laudable enough in having the potential to reduce the need for insecticides, but there is a potential problem with this. In tests carried out with modified potatoes that had a gene to produce a chemical naturally found in snowdrops which interferes with insect digestion, the number of aphids on the potatoes was markedly reduced. An unforeseen result, however, was an effect on the lifespan of an important aphid predator, the ladybird. Female ladybirds eating aphids fed on the modified potato plants were found to have their lifespan halved and males also suffered reduction in lifespan.

Natural genetic engineering

A possible, but as yet unproven, case of natural genetic engineering has been described in bacteria. It must be remembered that this is not a proven case, but it might be true. Demonstrating the truth of a past event, such as this one, is virtually impossible. So all we can do is review the evidence and try to make sense of it.

In 1885, while working in Munich, the German physician Theodor Escherich made the first ever description of a now very well-known experimental bacterium called *Escherischia coli*, commonly just referred to as *E. coli*. Over the years *E. coli* has been regarded as a benign and, indeed, sometimes a useful organism. It is found as a normal part of the benign flora of the human gut and does not usually cause any problems. However, in the early 1980s an outbreak of food poisoning was linked to a particular strain called *E. coli 0157*, so named because it has a particular cell surface protein called 0157 by which it can be readily identified. Now when earlier preserved samples were investigated by the Public Health Laboratory Service and cases of gastroenteritis from 1978 to 1982 were reviewed, it was found that there was just a single case where this strain of *E. coli* was responsible. However, it now seems that most, if not all, of the most severe life-threatening forms of gastroenteritis associated with *E. coli*, are, in fact, caused by *E. coli 0157*.

The distinguishing feature of *0157* is that it makes two different toxins not normally associated with strains of *E. coli*, but which are very similar to toxins made by other pathogenic bacteria responsible for severe gastric upsets. It is this unusual similarity in toxins that has led to the hypothesis that *E. coli* has managed to get hold of DNA coding for the toxin from a different organism, thereby changing itself from benign to pathogenic. If it is true, this would be an example of natural genetic engineering having taken place this century.

Although it might not seem obvious, what *E. coli 0157* did for itself is very similar to what genetic engineers can now do for other organisms. It appears to have shifted either single genes, or groups of genes, between different species.

Another case where toxin production is thought to have been introduced into a foreign organism, rather than being an original aspect of it, is cholera. If this event did take place then we have no method of knowing when it happened. Cholera has had a bad press over the years and quite rightly so. This is in no small part due to the work of Dr John Snow, nineteenth-century pioneer of public health. If you visit Broadwick Street, in London, you can see a plaque at the side of the road commemorating the events that resulted in the first epidemiological study of a disease. It all started in 1854 when there was an outbreak of cholera in central London. John Snow managed to demonstrate that it was the contamination of drinking water by sewage at a single public water pump that had caused the problem.

He persuaded the council to remove the handle from this water pump on Broad Street from which the contaminated water was being drawn, bringing a dramatic reduction in the rate of infection.

Cholera is caused by a bacterium called *Vibrio cholerae*. But this organism was not always the killer that it is today. It is thought that at some time before the great cholera pandemic of 1817, a virus, somewhat enigmatically called CTX, invaded what was until then probably a fairly benign *Vibrio cholerae*. This introduced a gene into the cholera organism which codes for a toxin that transformed the bacterium into a pathogenic organism and introduced the world to the appalling disease we know today.

Malaria and sleeping sickness

With the increasing sophistication of methods used to investigate pathogenic organisms, two very interesting examples in which microorganisms were found to be able to use techniques similar to those used by genetic engineers were revealed. In these cases the natural genetic manipulation of the microorganism is used to evade the immune system of the host. The complex host–parasite relationship of chronic infections tends to revolve around the immune system and the parasite's ability to either subvert the process or dodge it all together. These two examples are malaria and sleeping sickness.

Malaria is a world-wide disease; at any one time there are about 300 million people infected with the parasite that causes the disease, *Plasmodium*. This is a parasite of red blood cells that actually lives inside the cell and causes a serious, life-threatening illness. About one million people die each year from malaria, 90% of whom are in Africa. A very different parasite, but with a similar technique for immune system evasion, is *Trypanosoma*, which is the cause not only of sleeping sickness in humans but also of Chagas' disease, which is a similar disease of livestock. Trypanosomes can be up to five times longer than the diameter of a red blood cell. Both of these organisms – plasmodium and trypanosomes – seem able to evade the immune system of the host, causing serious damage and resulting in very unpleasant diseases. But how is it that these relatively large infectious organisms, both very much larger than viruses, manage to avoid being wiped out by the immune system of the host?

There is no magic process involved. These organisms obey the same biological rules as every other organism and are controlled, like everything else in the universe, by the same rules of thermodynamics. So why do humans not develop resistance to these diseases just as pests develop resistance to pesticides? The answer to that lies in the fact that pesticides are essentially both simple and static, but these parasitic organisms are not. Let us look at the way an invading organism is spotted by the immune system and dealt with. Every cell has molecules called cell surface antigens. These antigens are recognized by the immune system if they are foreign to the body so that the cells carrying them can be destroyed by the action of antibodies. This is the process that causes graft/host rejection in transplanted organs if the match between the recipient and donor is not close enough.

In the case of the two invading parasites that cause malaria and sleeping sickness, as they divide and reproduce they both exhibit incredibly high rates of a very special and unusual genetic process. This moves a copy of the gene coding for the surface protein from a position somewhere along the chromosome where it is not used, to a position at the end of a chromosome, where it is. This process of moving genetic material around a genome is called translocation. Every time the parasites divide and reproduce, some of them will have translocated a different gene to the expression site at the end of one of their chromosomes. In effect they have changed their surface coat. As it is the surface coat which is recognized, all the parasites with the old-style surface coat will eventually be destroyed by the immunological system as it recognizes them as alien and increases production of the relevant antibodies. But it will take time for the body to recognize those which have changed their surface antigenic coat. By the time this has happened the new, antigenically different parasites have multiplied alarmingly, and some will have changed their coat again.

So by manipulating their own genetics the parasites stay one jump ahead of the body's defences. Astute observers of this reasoning will have realized that as the parasites breed in the body and then are decimated except for a few changed individuals the disease should follow an 'up and down' course. Which is precisely what is seen with these relapsing fevers. The cyclical nature of the disease is more obvious in the case of sleeping sickness because the life cycle of *Trypanosoma* is more straightforward than that of the malaria parasite.

Despite all the chemical weaponry available, from insecticides to combat the insect vectors to anti-malarial and anti-trypanosome drugs, it has not so far been possible to eradicate these diseases. In truth, the best way of avoiding the problems of sleeping sickness and malaria is not to get bitten; insect repellent and mosquito nets really are essential in some areas of the world.

The invasion of Spartina grass

The largest-scale genetic manipulation which can be carried out, either in nature or in the laboratory, is the replication or deletion of a whole genome. This change in an organism's DNA content usually involves multiplication, or reduction, of the entire complement of chromosomes, rather than small pieces of DNA. Such large-scale changes do occasionally turn up naturally, in fact probably more often than we would imagine. The problem is spotting these events when they have happened, which is not always an easy task. One such example which has been recorded quite recently is the invasion of the waters of the south coast of the Britain by a natural hybrid of grass.

Basically two species of coastal grass – *Spartina alterniflora*, which has 70 chromosomes, and *Spartina stricta*, which has 56 chromosomes – cross-fertilized. They were originally from two separate and distinct geographical areas, which is probably why hybridization does not seem to have happened earlier, but having been brought together by human activity, they were given the opportunity to cross. At the end of the eighteenth century a new species of grass, *Spartina townsendii*, was discovered both on the south coast of England and on the adjacent coast of France. In the following decades it became progressively more common on both sides of the channel, becoming a nuisance as it invaded salt marshes and mudflats. This new species has 126 chromosomes, which is the exact number found by adding together the total complement of the two original species, and just to prove the point that this was how it originated, the whole process has been repeated on experimental plots by making artificial crosses between the two species.

Combating infectious diseases, new and old

So far we have looked at the results of naturally occurring examples of 'genetic engineering', and there are very many more, in all sorts of different organisms. A change that would cause a very little problem to a simple organism becomes much more disruptive in more complex organisms, so only very small alterations to a genome can be tolerated before the results become deleterious. Many people are concerned about the results of human activities on the environment, and that includes the environment of the genome within the cell. But natural processes can be quite as alarming. If it is true that virulent diseases can arise from a natural process of interchange of DNA coding for toxins, as is thought to have happened with cholera, that should be a genuine concern. We cannot possibly know where or when it might happen next; a new and potentially lethal infectious disease could appear at any time, apparently spontaneously.

Of course, now that we have more understanding of what has happened in nature we should be able to use the information to improve our ability to deal with pathogenic bacteria. One thing we do know for sure is that it is not the toxin against which we have to direct our activities, although this can be done, it is the organism producing the toxin. More precisely, it is the antigenic surface of the invader that the body recognizes and uses to produce the defensive antibodies. Stimulating this response is the basis of all vaccinations, from vaccination itself, which is against *Vaccinia,* the smallpox virus, through to polio and all those diseases which in the West are no longer regarded as major problems.

The idea of vaccination is simple: take a disease organism and kill it, or at least treat it so that it is not infective. Give that to a healthy individual and hope that the body produces antibodies to the antigenic part of the organism. If this process works then if that person ever comes into contact with the disease they are ready and able to combat the invasion. But sometimes the system does not work so well and vaccination against an organism has little, or no, value. Unfortunately cholera is just such a case, but there might be a way around this.

What if, say, we could remove the gene that codes for the toxin, but retain the rest of the genome? This would effectively stop the pathogenic part of the invader's activity, but still allow the body to react to it, thereby inducing some degree of immunization to the normal pathogen. We could, by this sneaky approach, simultaneously disarm the cholera bacterium and allow the body to handle further attacks by normal, pathogenic, versions of cholera.

This technique has been pursued successfully by a company in Switzerland, but has provided us with a difficult situation. The point here is that different countries have different rules regarding the control and use of genetically engineered bacteria. The new oral vaccine created in Switzerland against cholera was being tested on volunteers at Leiden in The Netherlands. The testing had been proceeding well when it was discovered that because it was a genetically modified organism special permission was required for its import into the European Union from Switzerland. This permission had not been granted. It was a bit late by then, however, because the volunteers were excreting the modified bacterium in their faeces, so the organism was now probably in the sewers and the North Sea. Besides which, since the oral vaccine was past the clinical trials stage in both Switzerland and Canada, it was in use in these two countries and would no doubt have already travelled around the world on passenger aircraft without any of the regulatory authorities being aware it! Bacteria have no need for passports and do not honour our political boundaries.

Somaclonal variation

When it comes to genetic manipulation of plants, different techniques are available from those used in microorganisms. One of these is called somaclonal variation. This was originally described as a method of inducing novel sources of variation among food plants. The technique is remarkably straightforward in principle, but the results, depending as they do on an element of chance, can be rather unpredictable. What happens is that clumps of plant cells are cultured in an artificial medium, during which time they undergo variation at a much higher rate than would be normally found associated with spontaneous mutations. Precisely why this rate should be higher is unknown, but it is probably a simple association with the totally artificial situation in which the cells find themselves. They are not under the overall control they would experience in the whole plant. Once cultured, each individual cell can be induced by the application of plant hormones to form a complete plant. This means that if a single cell has undergone a change, then when it develops into a complete plant, every cell will contain the same alteration to its genes.

Although, as I have said, the results are a bit random, occasionally a genetic change will turn up which gives us a new, and sometimes improved, food plant. It is interesting to note that for some reason this technique really only works well in species that have been subject to intensive breeding using traditional methods.

Commercial varieties that have been created using somoclonal variation include crunchier varieties of celery and carrot sold under the label of Vegisnax, a marketing name that I would have thought would put people off, rather than increase sales. There are many such examples but there is one to which, as an occasional cook, I give my seal of approval. This is a yellow bell pepper named Bell Sweet which is, unlike older varieties, almost seedless.

The search for ways to reduce seed production in crops where the seeds are not themselves the crop has become of immense importance to commercial growers. Recently it has become possible, using tobacco plants as an experimental model, to show that a gene can be switched on during seed development which eradicates the cells that make up the seed. This makes use of a gene which is present quite naturally and is normally used by the plant to destroy unwanted cells. By controlling its expression it becomes possible to stop seed production in the final crop. It is hoped that this can be used to control seed production in citrus fruit such as oranges. Although in this particular case it must have escaped the attention of the researchers that navel oranges are already grown that have no seeds.

The production of seedless crops has enormous potential for consumers, but probably of greater importance is that growers of seedless crops, especially in the third world, will become hostages to commerce. They will lose the option of growing crops from seed saved from last year's crop and will be forced to buy new seed every year. This is not just a money issue either, biodiversity will also be impacted. Traditional methods in which seed is kept back for the following year's planting allow slow changes to occur in the plants, maintaining the diversity of the crop.

Techniques of gene manipulation

When it comes to highly sophisticated techniques of genome manipulation, the use of these on mammals, including humans, opens the way to gene therapy and all that this implies for medical science. In these techniques a single additional gene is inserted into a nucleus. If all goes well, this is then integrated and expressed in the genome. Although it is only relatively recently that it has been possible to clone a single gene, this is the first step in being able to insert a gene into a cell nucleus. But how can we actually get this gene into a cell, remembering that the gene is tiny, invisible, and only just detectable with very sophisticated biochemical methods? Too small to handle, we still have to get it into the right place.

Depending on the type of organism involved, plant or animal, the method of getting the gene into the cell varies. Although we do not want to just produce a list of methods, after all this is not a technical manual, some idea of how it can be done is really quite fascinating. The methods available fall into three broad categories, with some of the them sounding like science fiction.

Let us start with the methods of physically penetrating the cell. These include microinjection and microprojectiles. The latter is, perhaps, rather more extraordinary. Very small tungsten pellets, of between 1 and 4 millionths of a metre in diameter, are coated with the necessary DNA and literally fired at the cells. This technique was originally designed for use on plant cells where the cell wall is far tougher and more robust than the membrane surrounding animal cells, but is now being developed for use with mammalian cells as well.

What microprojectiles do is simply introduce a piece of DNA into a cell, which is left to either become incorporated into the genome, or not, a process entirely dependent on chance alone. Similarly, a technique called electroporation also just dumps the DNA into the cell and gives it the chance to become incorporated and used by the cell. In this technique the cells are suspended in a medium containing the DNA to be introduced and the cell membrane is partially damaged by an electric field. This causes small pores to develop in the cell membrane, through which the DNA can pass before they reseal.

An alternative method is the use of a small vesicle called a liposome. A liposome is an artificial, hollow vessel made of essentially the same material as an animal cell membrane, a lipid bilayer. Because liposomes are fatty globules, similar to the cell, when they come into contact with a cell membrane they are sometimes able to fuse with it. Consequently, anything contained in the artificial liposome interior will be expelled into the cell. Putting a sequence of DNA into a liposome in the form of a gene which you want expressed gives it a fighting chance of entering the cell and being integrated into the genome, at which point expression might start.

As you can imagine, with all of these techniques the chance of getting the DNA into a cell is small, as is the chance of integration in the nucleus, and the chance that the gene will be in a suitable position to be expressed. Considering that all these factors seem to be very hit or miss, dependent as they are on a series of randomly directed chance events, it is no wonder that other methods have been investigated in an attempt to improve the take-up of genetic material.

A popular line of investigation revolves around the observation that some organisms already seem able to do the job scientists have been striving to achieve for many years. Some viruses are able to invade cells and integrate their DNA into the host genome, thereby hijacking the cell's metabolism and creating more and more viruses. This, of course, is the basis of viral disease. It would appear that for every cell type that is targeted for gene therapy with artificial DNA it may be necessary to have a specific viral vector. Concern has been expressed about this method of introducing DNA into cells. It is feared that it could activate genes that should not be activated, such as oncogenes which can be a cause of cancer. It would be equally damaging if the viral vector were to infect a completely different species that it was never designed to meet. Such a situation would rightly cause worries.

In Australia a virus has been modified with a gene for the outer coat of the eggs of mice. The result is that infected mice are rendered infertile by producing antibodies against their own eggs. This work was carried out with the long-term aim of stopping the intermittent plagues of mice that occur in Australia. But the announcement that this was going on was greeted with alarm. The fear is that if the virus infects another species with a similar enough egg protein, they too, will be rendered infertile. It would be bad enough if this was a rare animal, but if the same virus could mutate, it might be able to infect humans, with devastating results.

For many years it has been thought that animal viruses are just not able to infect humans, but recent problems with bovine spongiform encephalopathy (BSE) being linked to a similar disease of humans have thrown doubt on this assumption. The suggestion that engineered animals, such as pigs, might be produced specifically to supply organs for human transplants, xenotransplants, is similarly in doubt for these same reasons of safety.

When it comes to plants, a similar approach to introducing DNA into cells has been used quite successfully, not using viruses this time but bacteria. Some bacteria of the genus *Agrobacterium* occur naturally in the soil and have the ability to invade plants at sites of cell damage. Once associated with the damaged area a small plasmid, a circular chromosome which replicates independently of the rest of the bacterial genome, can induce growth of a gall. This is done by the tumour-inducing plasmid becoming incorporated into the genome of the host plant and its genes being expressed. By engineering these plasmids in the laboratory to carry wanted genes they can be used to induce desirable changes in the infected plant.

All these methods of gene manipulation have problems and all tend to have a poor performance; most cells are not affected, and those that are tend not to express the gene. The major question associated with this technology is why bother? Well, there are many different reasons, such as the possibility of gene therapy, the production of pest-resistant food crops and the production of useful products by expressing a gene in an alien species.

Gene therapy in species other than humans

It is clear that gene therapy in humans is desirable, especially for those hideous diseases that can cause such distress and heartbreak. It may not be quite so obvious in the case of pest-resistant food crops, and perhaps even more questionable is the use of biotechnology with an unproven safety record for purely spurious reasons. How would we feel, for example, about the use of these techniques to create a resistant species just for the aesthetic pleasure of humanity – perhaps a garden flower that could withstand a specific herbicide? Perhaps we need to look closely at our reasons for using genetic engineering.

Gene therapy is not necessarily confined to animal diseases. Currently attempts are being made to develop a variety of elm tree resistant to Dutch elm disease. This devastating disease, first recorded in The Netherlands in 1917 and imported to Britain in infected logs in the late 1920s, has been responsible for a staggering loss of more than 25 million trees in Great Britain alone since the 1960s. This represents in excess of 80% of the country's elm trees.

So why should we put resources, both intellectual and financial, into supporting a search for a method of saving a tree? Maybe we are a bit more altruistic than we imagine. There are, of course, a few groups of people who depend on elm trees, or did depend on elm trees, for a living; boat builders, for example. But even they could not, in a hard economic climate, justify all this expenditure simply for the benefit of a few individuals. Maybe we should not try to justify a search such as this in commercial terms. Perhaps we should just say 'it is worth doing because an elm tree is worth saving'. Without trees such as these we would all be impoverished.

Dutch elm disease is caused by a complex association between three species, a beetle, a fungus and a tree. Beetles of the genus *Scolytus* carry with them a small inoculum of fungus, essential for the survival of their young, called *Ophiostoma novo-ulmi*. When the beetles lay their eggs under the bark of a tree they leave some of the fungal spores behind. These start to grow, adding considerably to the nutrients available to the developing beetle larva. Unfortunately, as the fungus grows and develops it infiltrates the transport system of the tree, blocking up the water and nutrient channels, and ultimately killing the tree.

One control method that has been considered is to use the bacterium *Agrobacterium* to insert a gene into cloned elm tree cells, conferring resistance to the fungus. The *Scolytus* beetle has been around for a long time, and it is not primarily this that causes the problem. It is thought that a new fungal strain has evolved and this is why the disease has become so destructive.

The aim is to control the spread of Dutch elm disease, although there are some who would debate whether we have the right to intervene in this whole process. I would like to point out that the fungus probably did not originate out of thin air, it was almost certainly introduced from an area where it is a normal part of the ecosystem, arriving as an unwanted guest via the international trade in wood. If it was the human hand that introduced it, it is probably right that the human hand should be involved in its control.

Recently a new approach to this problem has been reported. It has proved possible to extract a protein from a benign relative of the Dutch elm disease fungus which can be applied via a small hole drilled through the bark into an elm tree. Within a few hours the elm tree produces anti-fungal agents which can protect it from fungal attack. Protection comes not in the form of antibodies but by the tree increasing the thickness of cell walls in infected areas, thereby sealing the fungus in so it cannot escape and block up the vital transport system of the tree.

Genetically modified food crops

When it comes to single gene insertions there are lots of products now potentially available, but few so far accepted by consumers. One such example is the FlavrSavr^{TM} tomato. This product has a much longer shelf life than a standard tomato, but the political feeling is that it should only be on sale in Europe as a tomato product, puréed and processed to reduce the risk of accidental gene transfer.

When fruit ripens it has a tendency to become soft and mushy before it starts to rot. This does not happen by magic, it is caused by a gene in the ripening plant which produces an enzyme called polygalactouronase. This enzyme breaks down pectin in the cell walls, allowing the fruit to lose its structure and collapse – a process called autolysis. The rotting, which soon follows, is carried out by bacteria and fungi feeding on the remains.

In order to slow down this natural process and produce a tomato that stays fresh and firm for longer, scientists have developed a method of preventing autolysis by blocking the production of the polygalactouronase enzyme. They have done this by inserting what is called an 'antisense' version of the polygalactouronase gene into the tomato plant. Antisense genes are genes that are effectively backwards. As we described in Chapter 2, once a gene is activated, the first process towards expression is transcription, a simple copying process, into RNA which can then be translated into a protein. In the modified tomato, when RNA is transcribed from both sense and antisense genes, instead of being translated into protein they bind together because they are complementary. Just like a lock and key, they fit together perfectly. Binding between RNAs produced from sense and antisense genes is dependent only on chemistry but it results in neither of them being available for translation into protein. This means that the production of the protein – the autolysis enzyme in this case – is blocked.

This remarkable work was carried out by an international group from Calgene in the USA, Nottingham University in the UK, Zeneca, also in the UK, and the Campbells Soup company. This interest by a food company indicates just how much commercial potential there is in engineered food crops.

Some consumers are less concerned with how their food is produced, whether it is genetically engineered or not, than what it tastes like. It is important not to get carried away by the technology alone and to get bogged down in arguments for and against. Taste and nutrient content are important when we are talking about food, but these are always subject to vagaries of personal preference. After all, iceberg lettuce was marketed on texture alone, but in terms of taste many people think it falls short of older varieties like Webb's wonder and cos.

When the first genetically engineered food was introduced in the UK in 1990 it was with almost no comment that a Dutch strain of baker's yeast was sanctioned for use. This yeast caused bread dough to rise more quickly because it had extra copies of one of its own genes and could thus digest malt sugars more efficiently. Since then all sorts of urban myths have arisen about genetically modified foods, largely because of a proliferation of incorrect information disseminated by the media. Such misinformation even extends to saying that genetically modified is the same as transgenic, which it most certainly is not.

Any alteration of a plant or animal genome can be called genetically modified – at one extreme one could include domestic dogs as genetically modified. Transgenic modification is quite different. This is the introduction of a foreign gene into a plant or animal. Thus the FlavrSavr tomato described earlier is genetically modified but it is not transgenic. It is produced by rearranging the plant's own genome and is thus probably quite safe. Questions of safety generally only arise with transgenic products.

One of the most widely grown transgenic crops is herbicide-resistant soya. By introducing a bacterial gene into soya bean plants it has become possible to make the crop tolerant to a weedkiller which would otherwise kill it. The result is that weed control for the farmer is much simpler and the herbicide can be used indiscriminately without any danger of damaging the crop. Another transgenic crop is a potato that has been modified to produce its own insecticide to reduce damage by aphids. In this particular case questions have been raised as to the ecological safety of the potato because ladybirds, the natural predators of aphids, seem to be adversely affected by eating the aphids that have fed on the modified potato plants.

Genetic engineering in animals

Understandably, the genetic engineering of animals raises many ethical questions and generates much debate. Most genetic engineering involving animals, and certainly all of the early work, was geared towards inserting a single gene into a species. This research was fuelled by the need to search for new methods of producing rare or expensive proteins for use in human medicine.

The production of novel gene products requires knowledge of the sequence of DNA in a gene. Ideally, both for production of transgenic species and also for gene therapy, the introduced gene should be carried on a chromosome that is not only stably inherited through generations but also originated from the recipient species. In 1997 a first step in this direction was made when the very first artificial human chromosome was constructed. This genetic milestone is a good demonstration of how the differences between classical genetics, molecular biology and cell biology are becoming ever more blurred. In order to produce an artificial chromosome we need not only DNA but also protein. Such is the robustness of chromosomes, it would appear that once set in motion and in the cell the DNA becomes a seed for a self-assembly system. This finally results in a complex structure which we know as a chromosome.

There are three fundamental parts to a chromosome, all of which are essential for its normal function and stability. The first of these is the centromere, the central point of the chromosome. The centromere controls the division of the genetic material during cell division and is of fundamental importance to inheritance. If there was no centromere, the chromosomes could not divide during cell division, and the daughter cells would end up with all manner of different amounts of DNA. The second important components of chromosomes are called telomeres. These are the end-caps of chromosomes and are important in maintaining their stability.

Because of the way in which DNA is replicated, without a very specific mechanism to retain the integrity of the chromosome it would become progressively shorter at each cell division. Every time DNA in the chromosomes undergoes replication, the enzyme that does this stops short of the end. It cannot go right to the end of the chromosome because if it did it would fall off the DNA strand. This means that there is always a short piece of DNA that is not replicated and the chromosomes would become shorter and shorter and soon lose their integrity after each division if this was not compensated for in some way. It is the job of the telomeres to perform this compensation and thus to maintain the integrity of the chromosomes. The third important part of a chromosome is the stretch of DNA that carries the gene that is of interest. This contains the information for manufacturing the protein that we want to introduce into a cell.

Once again, the process involved in introducing a foreign gene into an animal cell is relatively simple to describe and straightforward in principle, but requires great skill and dexterity of both hand and mind to execute. A popular species for the production of transgenic animals is the sheep. By introducing a human gene into a newly fertilized sheep egg, quite literally injecting the DNA into the nucleus using a very fine glass needle, and then implanting the egg in a ewe, it is possible to produce a lamb that will express the gene.

This is clearly a very hit and miss process: not all the eggs will integrate the foreign gene into their chromosomes, and of those that do take up the gene not all will be viable. Then you have to allow for all the eggs that fail at the implantation or embryo stage – not all will reach term. Finally, just because the lamb is born, it does not mean that it will necessarily express the gene that was originally implanted and produce the protein of interest.

Once a lamb is born it can be screened to see whether expression of the gene is taking place. If it is, the question remains as to how the protein product can be extracted from the cells. In the case of lactating animals it is usually simplest to just milk the animal and extract the protein from the milk.

Because of the unpredictability of this method of gene introduction and the difficulty in achieving the desired result, another technique has been developed. This contains an additional step and uses a different cell type, embryonic stem cells. These cells, which are derived from embryos before implantation, can be grown in a Petri dish in the laboratory. Once the required gene has been introduced, the cells can be grown on in tissue culture and tested for expression of the gene. If the experiment has been successful and the gene is being transcribed with the correct product, the nucleus can be removed from one of the cultured cells and introduced into the nucleus of a cell within an embryo. Once born, the offspring will have some cells producing the wanted protein and some not.

The next generation, however, will be divided into non-expressers and complete expressers. This is because at the time of meiosis some of the sperm or ova will have the gene and some will not, there are no half-way houses at this point. The result of this is that complete expressers are true transgenics, carrying a foreign gene in all their cells, and often producing the required protein in all their cells as well.

The first successful transgenic sheep was produced in 1989. This animal produced the anti-haemophilic factor IX in her milk, and although the amounts recovered were small, researchers were able to see the immense potential for the production of this vital blood-clotting factor in this way. It was clear that this method of producing factor IX could make a huge impact in the treatment of haemophilia.

Perhaps the most successful transgenic sheep, and certainly the most famous, was Tracey. She produced about 1.5kg of a protein called alpha1-antitrypsin in her milk annually. In humans, deficiency of this protein is associated with a progressive lung disorder called emphysema, and treatment of this condition requires about 200g of alpha1-antitrypsin per patient per year. Clearly transgenic production of it is a step in the right direction, remembering that it still has to be extracted from the milk in a useable form. Ideally, the production of a transgenic cow would be a great boon as cows can produce 10,000 litres of milk in a year, while sheep at best can barely manage 800 litres.

Dolly the sheep

In the late 1990s another sheep, this time called Dolly, created an enormous stir, both within the scientific community and in the popular press. The production of Dolly apparently took many people by surprise; in fact, it had a huge effect in highlighting the lack of understanding of the general public about progress in genetics and gene research.

Ever since the earliest experiments in embryology were conducted, it had been thought that it was not possible for a mammalian nucleus to be converted back from a fully differentiated state to one that could produce a complete organism. In some ways this is a naive line of thought; after all, if it happens in nature there is no intrinsic reason why it could not be accomplished in the laboratory. What has always given concern is that it has never previously been possible to reverse the process of differentiation – the process of change as a cell develops into its final form. All cells have the same genes, but some are turned on and some are turned off. Once committed, a cell uses only some of the total genome with the result that cells become muscle cells or nerve cells, but never both. As a cell ages the DNA content does not alter, neither does the sequence; what does change is the number and types of genes which are switched on and off. Some genes are vital for early development and are normally then switched off completely, but if they are accidentally switched on at the wrong time the resultant cell proliferation is a cancer. It follows that if this switching on can happen accidentally in nature, then it should be possible to make the manipulation ourselves.

What happened in the case of Dolly was that a team at Edinburgh, made up from members of the Roslin Institute and PPL Therapeutics, managed to clone a sheep from a single cell taken from the udder of a ewe that was already six years old. This was the first time that a mammal had been cloned from a nucleus of a non-reproductive cell and it was a staggering scientific achievement. Dolly can realistically be described as the most amazing animal ever born.

However, let us put this research into perspective. Dolly did not come completely out of the blue, although the press of the day would have you believe so. In 1996, one year earlier, the Roslin Institute had cloned Morag and Megan. These were two lambs, but instead of being cloned from cells of a non-reproductive origin, they were cloned from cells taken from an already fertilized embryo, which was then split and two cells put into a foster mother to develop and be born. This means that there is an historical precedent for Dolly. But what is it that has made the cloning of an animal so long in arriving? We have, after all, been producing new humans since our species evolved. Admittedly this is not a cloning process, indeed, it is far more complicated and produced the extraordinary variety of human beings which we now see. But for those wise enough to question the laboratory technology the immediate response is often 'why bother?' I suppose the answer could be simply scientific curiosity, but I bet there will be little call for production of human clones, although short-sighted individuals may wish it so. The reason for this will become clear as we discuss how cloning is achieved, but the ethical questions, rather than the scientific ones, will be dealt with in the next chapter.

The techniques of cloning

The process of cloning starts with the realization that a cell from an adult is not normally able to reproduce indefinitely. Without a mechanism that allows for the regeneration of chromosomes the cell will eventually run out of the very DNA that defines the cell itself. This probably requires some explanation. After all, we, along with all other animals, have been reproducing for generations, and as yet we have not run out of chromosomes.

When a chromosome is replicated, one of two courses of events occurs. Either the whole sequence, end to end, is reproduced, or each time the chromosome is replicated it gets shorter. Unfortunately, under normal conditions it is the second of these two conditions that prevails. Enzymes which replicate DNA, called DNA polymerases, use one half of the double helix as a template for the generation of a complementary strand. Strangely, replication of

DNA is unidirectional; one strand replicates in one direction and the other strand replicates in the opposite direction. Now, DNA polymerase is a very large molecule, so when it sits on DNA to begin replication it covers part of the DNA to be replicated. Think of it as a train sitting on a railway line: as it moves forward it can replace the track in front of it, but the section of track it started on is not replaced. This means that on each DNA strand the bit where the enzyme starts work will have a section which is simply not replicated. This section of DNA is called a telomere and cannot be copied by the standard DNA polymerase.

Rather surprisingly, all the telomeres that have been sequenced throughout the animal kingdom have been found to be very similar in sequence. Although not identical in all cases, they follow a repeated DNA sequence based on TTAGGG. This base sequence is repeated many times and fits in with a special template in the DNA polymerase molecule which enables the cell to actually add extra DNA to the end of the chromosome. This might seem drastic, after all, just as we do not want ever shrinking chromosomes neither do we want ever growing ones. Under normal conditions chromosome extension at the telomeres is confined to very early embryology; the rest of the lifetime of the organism sees a progressive shortening of the chromosome at each cell division. It would seem that cancers are the exception to this, otherwise of course the cells would die rather than just endlessly proliferating.

This chromosome shortening during life has been associated with a curious observation made by L. Hayflick during the 1960s, and now called the Hayflick phenomenon. What he recognized was that when animal cells are grown artificially in culture, they only seem capable of a pre-programmed number of cell divisions. Using human cell cultures, Hayflick showed that cells from four-month-old embryos doubled about 50 times before stopping, cells from 20-year-olds doubled about 30 times, and those from 30-year-olds only 20 times. Even when embryo cells are allowed to double 20 times and are then frozen in suspended animation for several years, once defrosted they still only carry on doubling up to the Hayflick limit. It is possible that this limit is delineated by the destruction of the telomeres and that the slow erosion of telomeres is what eventually stops cell division. So what does this have to do with Dolly? Simple, the ewe from which the donor cell was taken was six years old and it is therefore probable that Dolly will have a much reduced lifespan. Given that Dolly's telomeres will be those of a six-year-old, it seems reasonable to suppose that she will have the lifespan of a six-year-old.

Fortunately, there is a reason to think that this untimely death will not occur, because of the methods used in Dolly's cloning. The process that was used involves removing the nucleus from an oocyte, that is an egg cell, and replacing it with the nucleus from a mature, differentiated cell. This was not a one-off, single shot event; Dolly was the only birth from a total of 277 fusions that were attempted. The DNA in the nucleus was added to the egg by subjecting the two cells to an electric current, fusing their membranes and resulting in the nucleus of one cell with the cytoplasm of two. Before implanting the newly formed cell in the host mother, it was cultured in the laboratory for about a week until the early embryonic form called a blastocyst had developed. This developing mass of cells, grown so far *in vitro*, was then implanted in an animal receptive to pregnancy. Now because an undifferentiated cell, an egg, was used to house the nucleus, it can be assumed that the cytoplasm of this cell would contain enough of the necessary enzymes to resurrect its chromosome telomeres. If this happens then Dolly would be expected to have a normal lifespan. We shall see. Dolly was born in 1996 and is still alive at the time of writing, but as sheep normally live around twelve years, it will be some time before we know the outcome.

The implications of cloning

There has been a rash of prophecies regarding cloning, often associated with the fear that in the hands of unscrupulous tyrants, human cloning could be used to produce terrifying armies of super warriors or unlimited copies of powerful leaders. Perhaps less in the realms of science fiction there has been the suggestion that cloning could be used in animal conservation. For example, Barcelona City Council has put forward the idea that the very rare albino gorilla in

Barcelona Zoo should be cloned. This has been hotly denied by zoo staff, I should add.

Fortunately, there are several factors that militate against the possibility of doing any such thing, whether we would like to or not. Some of these are based on sound science and some, at this stage at least, are in part conjecture, but none the less valid for all that.

For starters there is the question of the mitochondrial DNA. When we speak of a clone we mean genetically identical, but to return to Dolly for a moment, there is at least one way in which Dolly is not genetically identical to her mother. In every cell there are two types of DNA, nuclear and mitochondrial. As we saw in Chapter 3, mitochondrial DNA is inherited only from the mother, which gives us the opportunity to trace a maternal line of descent by following the mitochondrial DNA, as was used in the case of the Russian royal family. Three different sheep were used in Dolly's production. There was the individual from which the nucleus came, there was the donor of the egg cell and there was the ewe that carried and gave birth to her. Now it seems reasonable to assume that Dolly's mitochondrial DNA came from the donor of the egg, not from the donor of the nucleus. If this is the case then Dolly is not, strictly speaking, identical to the donor of the nuclear DNA; she would be a nuclear clone, but not a complete clone of her donor DNA. This may seem a small point, but techniques of cloning have shown us the possibility of separating completely the effects of nuclear DNA and mitochondrial DNA from each other. This has considerable importance for some families where mitochondrial DNA, separate from nuclear DNA, has implications regarding inherited diseases.

Another aspect of cloning that is easily overlooked is that at the moment we cannot determine the part that nature has in the final outcome of an individual. The part played by nurture, our upbringing, might be far more significant than we imagine. This is a truly old argument, but what it comes down to is the question of whether your heritage, in the form of DNA, or your upbringing, nurture, has the greatest effect on who you are. Since we are saying 'who you are' we are touching on the very central essence of humanity and it would be wise for us to tread carefully. There can be few people, scientist or non-scientist, who have not considered this question. What is, after all, the contribution of your genes and your background to who you are?

It is clear that the impact of environment on the formation of personality cannot be ignored. With a difference of six years between Dolly and her progenitor, it may not be so obvious, but with humans I can assure you it would be. By the time some despot decided to clone himself his formative years would be long gone and they can never be reconstructed. Time moves on in human history. A human clone may be genetically identical to its progenitor, but will most certainly be a different person.

With regards to the cloning of livestock, it has been pointed out, and quite rightly so, that cloning reduces the genetic variation in a crop. This is true whether we are dealing with a plant or an animal, and plant monoculture has not proved the wonder it was originally thought to be. Biologists and even politicians are now realizing that there really might be a significant reason for biodiversity; it is not just a perverse attempt to confuse taxonomists.

Another possible use for cloning that is often suggested – usually in fiction – is the resurrection of long dead animal species or even humans. As we discussed in Chapter 4, it is not going to be possible to resurrect dinosaurs or any other extinct species using cloning. When it comes to cloning from dead material there is a problem. As soon as decay starts, the delicate balance of genetic material starts to break down. Although DNA is a relatively robust molecule in itself, the associated proteins are far more delicate. Once damaged, these proteins can no longer operate correctly, so at this stage just knowing a DNA sequence is not enough. Even knowing the complete DNA sequence of the human genome will not allow us to construct a clone; the secrets of the cells which go to make us up are still hidden.

Perhaps we will be able to construct a human clone one day, but why should we? Humanity thrives on variety and the usual methods of reproduction have not only served us well throughout time, but have a great deal to be said for them in terms of simple pleasure and social bonding.

EIGHT

Whose DNA is it anyway?

The control of information now and in the future

There is an unusual content to this chapter for several reasons, not least of which is that because while we might have a clear idea of the sorts of ethical questions that we would like to answer, that gallant band of philosophers who study ethics as a subject tend not to deal with specific issues. This is not to say that they cannot, or do not want to, but merely that they approach ethical questions rather like the rules of algebra. An attempt is made to produce formulaic rules of thought with which to guide the philosopher, or ethicist, to a satisfactory conclusion.

Unfortunately, for more modest thinkers this does not always help. Indeed, very often the rules of thought are not as you or I would wish them to be, that is all ending up with the same answer, but one school of thought tries to browbeat another with logic into accepting either their current dogma or point of view. There are, however, many interesting points to be gained from a theoretical consideration of ethics when applied to specific questions.

The impact of science on humanity

Nearly all advances in science have some impact on individuals and groups. Sometimes this impact will be negative in outcome, as in the threat posed by the development of nuclear weapons or biological warfare. And even where these do not directly injure a particular individual, the stress caused by the existence of the threat can be quite debilitating on its own. Other advances, like antibiotics and the supply of clean water, may be perceived as entirely positive. But there again, the production of better pharmaceuticals, including antibiotics, has required testing on animals, and this may be seen as untenable. This leaves us with a question that cannot be answered because it depends on the personal view of the person answering it. 'Can the ends ever justify the means?' Most would give a qualified yes to this question, even remembering that in 1991 there were just over 3 million animals used in experiments in Great Britain alone. Luckily this number has been steadily falling as *in vitro* techniques for testing improve.

It is for these reasons that we will start with a practical assessment of the ethical questions that arise from knowledge of DNA and see if we can resolve them satisfactorily. These are not only ethical questions, but also simple questions of safety. It is also useful to look briefly at the theoretical considerations of ethics. When these two are put together we should at least have a framework for approaching questions which arise about this work both now and in the future. Make no mistake about it, there are going to be many more questions as technology changes and becomes more sophisticated and this is bound to create fears for the future.

Trying to second guess the future is a road pitted with potholes, but we can look at where some scientists and politicians would like DNA research to take us and ask what are the issues which we will have to address, both as individuals and as a society, and how exactly is this technology going to affect me and my family?

Giving ourselves an informed base

Whenever questions about the use of genetics and DNA arise, fears are expressed. These fears may be based on misunderstanding or ignorance, but they are fears none the less. Without doubt there is more concern over genetics, genetic engineering and eugenics as a science and subject for legitimate debate than any since the advent of nuclear physics. Of course there is a fundamental difference between these two. It was quickly recognized that research into atomic physics had a huge potential for impacting national security. This meant that it was primarily driven by governments and the military. Research in genetics, on the other hand, was initially driven by its medical and commercial potential.

It has to be said that scientists working in genetics have acknowledged the concerns of the public and have taken a lead in hosting public meetings where questions can be asked and, where possible, answers given. Biologists and geneticists have become very aware of the need to communicate their science to a wider public, not just to justify their work, but to help people to understand what is being done in their name. Scientists do, after all, have a duty to tell people what research is being done in fundamental sciences since, unlike that in applied science, it is for the most part paid for out of the public purse. Admittedly the Wellcome Foundation, which is the largest non-governmental medical research charity in the world, does fund a great deal of research, but it is still governments which pay for the bulk of pure research. The Wellcome Foundation has, itself, organized meetings to help in the understanding of genetics. Nobody could deny the immense value of communication in this area, as the results of lack of understanding can be severe. Recognizing the importance of increasing the knowledge of science in the general public, the Royal Society has a well established Committee on the Public Understanding of Science.

When the Wellcome Foundation ran a meeting for the public, Professor Martin Bobrow of Cambridge University said, 'Technology only sets the limits on what is worth talking about – and there you are the experts.' This was a very astute observation; if society does not want genetic research, both pure and applied as it is done, then they should not only say so but also be able to stop it being carried out. But, and it is a big but, let us not make snap decisions based on inappropriate and inaccurate information, which is often gained second-hand from misinformed sources. It is here that the scientist not only is able to have an input, but must make sure that the informed decision of the public is made only from true information and solid data. It is not right in genetics, or any other subject for that matter, that attitudes are created by a blinkered vision that only allows selective perception of data – that is no way to run a society.

Without a doubt there has been a great deal of misuse of science over the ages and genetics is no exception to this. Unfortunately geneticists do not always get a good press and tend to be seen, I suppose, as the potential perpetrators of evil. It is indeed a fact that scientists themselves have not always been innocent and geneticists have sometimes encouraged ideas that we now find untenable. It is perhaps easy to judge with hindsight, but it is important that we should examine the past in order to put our science in perspective, bearing in mind that genetics is so new that mistakes tend to be recent rather than ancient.

The perspective of the past

Misrepresentation of all sorts of science has a long history. For example, the so-called scientists of the Pope managed to subvert astronomical observations for centuries before the truth of earth's orbit around the sun became acknowledged and it was only in the 1990s that the papacy officially acknowledged that the sun did not go around the earth! Science is not, and never has been, an activity carried out for the benefit of a few. It is an exciting and exhilarating activity which should be creating opportunity and advantage for everyone. To this end the interpretation of science should not be at the whim of politicians, even though that has happened in the past.

Genetics as a science can be said to have started in the nineteenth century, not with Gregor Mendel, although he was working during this period, but with a cousin of Charles Darwin called Francis Galton. It was Galton who came up with the premise that to a great degree intelligence and talent are inherited. In his book, *Hereditary Genius*, published in 1869, he developed his ideas on this subject and it was here that he first described a new science that dealt with breeding a better person, a science that he named eugenics. Simply put, his argument was that if we want better people then we should selectively breed them. Even as an idea, let alone a practical proposition, this would not be countenanced now, but at the time it seemed quite reasonable.

Unfortunately for the rest of humanity ideas such as these were taken up and promulgated by the ignorant for their own purposes.

One proponent was Victoria Woodhull in America who, in the late nineteenth century, published two books, *Stirpiculture, or the Scientific Propagation of the Human Race* in 1888 and *The Rapid Multiplication of the Unfit* in 1891, in which she stated that if superior people were wanted then they should be actively encouraged to breed. She also propounded the alternative: that the unfit must not breed. The readiness of the American people to accept these ideas can be gauged from the fact that this eccentric lady was the first woman nominated as a presidential candidate in the USA and that by 1930 some 24 states had passed legislation either requiring, or allowing, the sterilization of the 'feeble-minded'.

For many years thereafter both positive and negative eugenics were regarded as legitimate areas of speculation among those sufficiently well off to feel themselves in some way above the common herd. Positive eugenics is the process of preferential breeding of people with 'good' qualities, while negative eugenics is the opposite; persuading the 'unfit' not to breed. Many countries had Eugenic Societies, which existed to cater for the new science of genetics. Without exception they changed their names when eugenics developed a sinister and unacceptable reputation with the rise of fascism during the twentieth century.

Although I would not attempt to be an apologist for what has gone on in the past, it must be remembered that the hideous actions of the twentieth century which have masqueraded as science have almost without exception been the results of political machinations by power-hungry heads of countries.

It would not be productive to berate ourselves for the past; we can do nothing about what has happened, even if it was only a microsecond ago and regret is, after all, a rather negative emotion. Nevertheless, we can try to learn from history. However trite this statement seems, we have all experienced, read or heard about the stupidity of dogma when applied by one nation to another nation. So let us at least try to modify our attitudes towards scientific work, and in this context genetics in particular, especially in the light of the past. Do not ignore the possibilities, good or bad, but consider it a challenge to decide just where you want our future to lie in respect to science and technology.

National policies on genetics – the case of China

There seem to be many more troubled minds contemplating the possible misuse of genetics in our research institutes than many people would give credit for. Indeed, there are probably more than are contemplating the problem outside academic circles. It is no longer seen as acceptable just to 'do' genetics without accepting responsibility for the results. So let us move forward and look at the problems of both now and, if it is possible to guess such things, the future.

Unfortunately, the idiot ideas of eugenics have not yet been laid to rest. While most people would applaud our genetic diversity and recognize that there might well be a reason why we have so many carriers of genetic diseases in the population, sometimes the evidence is ignored and our genetic diversity compromised. During 1995 the government of China brought a law into operation which forbade couples diagnosed as having genetic diseases from marrying, unless they were sterilized or using long-term contraception. What this draconian measure fails to appreciate, besides the intrinsic rights of the individual, is that all these measures do is to cause pain and heartache. What they will certainly not do is reduce the incidence of genetic disease in the population. For one thing, some genetic disease genes, such as that responsible for Duchenne muscular dystrophy, have such a high mutation rate that it has been estimated that even routine testing of every individual for carrier status would not help in deciding who would be at risk of giving birth to an affected baby.

Unfortunately, there seems to be a tacit if not explicit belief that some genes are good and some genes are bad. Such ideas are flawed from the very outset. Take, for example, oncogenes. These genes have been known for some time to be implicated in the development of cancer, hence their name, but they are also found as normal parts of the human genome. Indeed, oncogenes have now been shown to be essential for early development; researchers have reported that if certain oncogenes are defective then the cell cannot divide. This means that although we can say that inappropriately switching on an oncogene may result in a cancer, we also know that without them we would not exist and that the eradication of oncogenes in an attempt to conquer cancer would eradicate us. Although an extreme example, this does serve to illustrate that arbitrary decisions about whether a gene is good or bad and therefore suitable for removal from the human genome can be flawed by a simple lack of knowledge.

The bottom line is that we should be concerned when a political regime apparently committed to eugenics gets started down the path of deciding what is or is not genetically good for its people. There can be no ultimate arbiter on this, except possibly the individuals who are on the receiving end of this behaviour. In traditional terms, governments do not have to answer to anybody but invading armies, although more recently political pressure has sometimes worked.

Had the Chinese government approached this problem in a different way we might not have faced this issue at all. Surely it is in the world's interest that better therapy and treatment for disease is produced, rather than risk reducing the genetic diversity of the population. These draconian measures will not only fail in their attempted aim to eradicate these conditions, they will also expose the population to the risks associated with a reduction in genetic diversity. Although there is a profound difference in cultural expectations between western society, based essentially around promoting the good of the individual, and Chinese culture, based on the good of society, it surprised many western geneticists when a survey revealed that more than 90% of Chinese geneticists supported the idea that parents both carrying the same defective gene should not be allowed to have children. For some, the even worse statistic was that less than 50% believed that a woman should have the right to decide whether to have an abortion or not. Since then it has become possible for a doctor in China to order a termination if it is believed that the foetus has a defect of a serious nature. We should be alarmed about this if for no other reason than that it directly affects one fifth of the human population of this planet, officially made up of 56 different ethnic groups.

This particular chapter of our history is made worse by the involvement of a group who should know better. A French company, by the name of Genset, has undertaken to test, anonymously, Chinese nationals for genetic diseases. It is easy to see the possibilities. Even if we think that a government is sincere in its protestations towards the sanctity of the individual, can we really be sure that a benign government today will never give way to a more underhand government tomorrow? If we do have any doubts about this, why not set up safeguards against it now? It would not be paranoia – after all, the same is true of long-term storage of nuclear residues, the risk is great, so safeguards have to be put in place. So there is no reason at all for not safeguarding our genetic data and heritage, which would mean that there is really no reason why the information gained from genetics should ever be misused. Under such conditions there may be very valid reasons why, if nations deny access to defence material, as they do now, they should not also deny access to genetic information, or perhaps genetic technology, to dubious regimes.

At first glance a possibly even more bizarre position on genetics is being taken by the government of Singapore. There they are encouraging wealthy families to have large numbers of children, an idea redolent of Lamarckism. This seems to imply a belief in the inheritance of initiative, good business sense and attitudes to risk. While some of these might be influenced by factors inherited through DNA, it is clear that the environment has an overriding influence. If wealthy families tend to have wealthy sons and daughters this is not a consequence of genetics but of opportunity. Once a self-made man has a family their potential for wealth and status immediately increases as well. This is inheritance, not in the 'loads of money' sense but in the sense of access to opportunities to enable an individual to achieve his or her maximum potential.

The risks of misuse

The questions discussed here about the long-term political control of genetic data are not raised in the context of conspiracy theory. They are raised with a view to the passage of time and our historical position. Let us continue the comparison with the case of nuclear waste, using logic to see what the implications are for us all and future generations. If nuclear waste requires storage for, say, 10,000 years, can we feel sure that safe storage systems are in place that will survive that long? Given that no civilization, anywhere in the world, is that old, it seems unlikely that any nation presiding over the waste dump will still be extant and responsible that far into the future. The simple answer is that we cannot be sure that nuclear waste will be stored safely until it is no longer hazardous. Similarly, we have no way of being sure that when politicians say that no misuse of genetic data will take place this will also hold good for all their successors.

It is, without doubt, impossible to correct mistakes of the past and, unfortunately, is it impossible to predict the future with certainty. Already we have an example of scientists, medics and members of the general population digging their collective heels in to stop a political initiative dealing with genetic data. On the 17 December 1998 the Icelandic Parliament passed a Bill by 37 votes to 20 to allow the accumulation of records from the whole of the population of Iceland, currently about 270,000. This would include medical and genetic data as well as genealogical data. It is the existence of the extremely well-kept genealogical records of Icelandic families that makes genetic data from this well-defined population potentially so very valuable. This database could be of immense value for researchers looking for the genes that cause disease. Under the Icelandic Bill the government would allow a single licensee to both collect and use the data. Although personal data are said to be unwanted, suspicions have been raised that it is simply not possible to put secure enough safeguards in place to prevent the misuse of this information in future. These concerns have so alarmed the scientists involved that they have, in large numbers, effectively walked away from the project.

What we might be able to do, however, is generate a society where use of genetic data by third parties is considered unacceptable. Similar changes have taken place in attitudes in other areas, a most noticeable one being slavery. Of course, slavery in the United Kingdom has only been outlawed for about two hundred years, so whether this attitude towards legal bondage will stand the test of a thousand years we have no way of knowing.

Xenotransplantation

Some issues in genetic research induce a squeamish sense of distaste in some people. It is important to remember that these feelings might well mislead us into responses that are so entirely personal that they have little to add to the debate. At the same time we should not be dogmatic in our answers to these questions, as very often the outcome is entirely personal. An example of an issue of this sort is the possibility of xenotransplants. These are transplants from one species to another, for example the transplanting of an organ from an animal to a human.

Before we look at the human aspects, let us add a little perspective by taking a general definition of xenotransplantation as the transplanting of material from one species to another. In the plant world, xenotransplantation has been used for many years to improve the growth of plants used in horticulture and agriculture. Most people who grow fruit trees or roses are familiar with the idea that a plant can be stock grafted onto an alien root to improve vigour and growth. This is an example of a xenotransplant. Although I appreciate that the idea of a human receiving an organ from, say, a pig is rather more emotive, it is important to keep a rational base to this discussion.

From a medical point of view, any form of transplantation is intrinsically difficult because of the problems of immunology. Even when using a transplant from the same species it is not possible to guarantee that the transplant and recipient will thrive. Thus it would seem even less likely that one from a different species should work. However, it might be possible to create immunologically compliant animals, and today research to this end is well on the way to success. The idea is to genetically engineer pigs to give them a gene which produces a protein that inhibits the immunological reaction responsible for recognizing and destroying the transplanted organ as alien. This technique is, admittedly, only in its infancy as a potential supplier of organs, but it is at this stage that questions should be asked regarding the ethics of this process, both for the donor animal and the recipient. It is very little use asking questions after the event; once it is possible to make viable xenotransplants it will be quite impossible to put the genie back in the bottle. Remember that there are approximately 150,000 people around the world waiting for suitable human donors for various organs. And it is not just 150,000 lives that are affected, the lives of all their families and friends are impacted as well.

There are two basic questions to be addressed here. The first I believe we have a right to ask, but the second I believe can only be asked with a great deal of caution. Let us ask them anyway, and see what we can work out. First, do we have the right to manipulate the genome of animals for our own uses? We each have to make up our mind on this issue. The other question, closely allied, is whether we have the right to breed animals for uses other than food, specifically as organ donors. My own view is that as we have been breeding animals for food for thousands of years it seems entirely consistent to use their genomes in this way. It is not really so different. Instead of using animals as suppliers of food, we are going to use them as direct suppliers of body parts. As I say, this is a personal opinion and we all have to make up our minds on these issues. Admittedly, some vegetarians may argue that we should not be using animals for food, let alone for body parts, but it cannot be denied that meat-eating does have a very long human tradition.

Of course, there are many other reasons for selective breeding of animals, besides food. Cart horses and racehorses did not develop by chance. Traditional methods of animal and plant breeding have manipulated the genome just as much as genetic engineers have, and although traditional breeders work on a longer time scale I do not think we should be too certain of the goings on of past breeders when developing new strains of animal or plant. Take, for example, the case of the bullmastiff and the Mexican hairless dogs. These both had a common ancestor in the form of some sort of wolf or wild dog, but long-term manipulations by breeders working on time scales of generations have resulted in radical changes. They are so far from their wild ancestors in appearance that it is difficult to imagine the resultant breeds being anything but distantly related to either their ancestors or each other. In evolutionary terms, of course, they are really quite closely related, although they can be viewed as having undergone highly accelerated evolution.

Plants have not fared any differently, and again it is human vanity that has driven the creation of so many new varieties, for some of which it is almost impossible to identify the original stock. This is probably more so in garden varieties than food crops, because garden plants are usually bred for their appearance. Of course, a difference between plants and animals is that it is a little more difficult to control the breeding of plants. They do have a tendency to be rather promiscuous with their genes on their own.

In summary, it is clear that alteration to the genomes of our crops and genetic manipulation of our animals has been going on for thousands of years and this is why I feel quite content that we have a right to manipulate the genomes of animals for our own uses. However, there is a difference when it comes to the production of donor animals. If we can recognize the gene in the donor species that causes rejection it is possible to introduce an alien, or blocking, gene so that when a donor organ is taken and put in another species, the organ survives. Not so many years ago it would have been inconceivable that inter-specific transplants would work because it was so difficult even to get transplants from one individual to another of the same species to work. It is this that gives us the ethical conundrum and I would suggest that this is an area of legitimate speculation, but that the next question is not.

Once an animal donor of organs or just replacement cells has been produced and the tissues made available, I would suggest that nobody has the right to interfere in the personal decision of the recipient whether to receive the transplant or not. Given the choice between receiving a kidney from a pig and thus the opportunity of a prolonged life, and the prospect of a painful death, I suspect that most people would opt for the former. Their need for survival would outweigh any qualms they had with the concept of xenotransplants. It would also be quite wrong for a fit and healthy person to take the moral high ground and to consign another person, the potential recipient, to death by taking away the opportunity offered by xenotransplantation. That would be neither fair nor moral. Indeed, it could be argued that since we have the ability to produce the animals, or soon will have, it would be just as wrong not to produce them as it would be to produce them but deny their usage.

I do not want this to appear as though I am suggesting an unreserved free for all when it comes to the use of our knowledge of DNA. In fact I would like to sound a cautionary note. It has been reported in a study from Guy's Hospital in London that genetic counsellors often do not just give information, but rather too often give advice and direction. In some cases counsellors were found to be saying what was 'best' for the clients, or telling them whether they had made the 'right' decision. Interestingly, the counsellors also tended to judge, consciously or subconsciously, whether clients needed more 'guidance' by virtue of their assessment of the socioeconomic status of the client.

Now, nobody would suggest either that the advice was wrong or that advice was incorrectly given; the counsellors were clearly trying their best to help the people they were dealing with. In this light the result of such a study is not so much a surprise, really. Giving advice, as everyone knows, is an activity fraught with problems. If you only give pure information on a subject it might, quite literally, be impossible for the receiver to make a sensible decision. What people often want and need is either guidance or reassurance for their decisions, not just raw data. But just how much interpretation of results and guidance should be given is a grey area. In genetics in particular, just giving factual evidence can result in more confusion if it is not understood. In the event it may be necessary for genetic counsellors to interpret the results of tests for clients, whether we like it or not.

Thomas Guy, the founder of Guy's Hospital

Interestingly, the founder of Guy's Hospital illustrates both the problems of using hindsight and ethical questions which have stayed with us through the years. Thomas Guy was a local lad; that is, he was born and brought up in the area of south London near London Bridge where Guy's Hospital would eventually become established. Born around 1644, the exact date being unknown, Guy is usually thought of as a philanthropist. As a young man he earned a living as a bookseller and publisher, later making a great deal of money from importing and then printing Bibles. He used some of this to set up the hospital that carries his name.

To look a little deeper into the financial success of this man, we should also know that he made a great fortune from selling out of South Sea shares. These were associated with a company which at one point in 1720, before its collapse, was supposedly so wealthy that it offered to pay off the National Debt of Great Britain! But most of the money in this system came from the traffic in slaves. When Guy died, he left Guy's Hospital an endowment to the sum of £219,449. The question is, should we now despise Guy's Hospital and press for it to be closed because of our moral perceptions of the past?

Projected backwards by nearly 250 years, our current perceptions of morals just do not work. Surely we should accept the past and try to make both the present and future better for us all.

Genetics and the insurance industry

So far we have looked at cases where it is possible to stand back when making our judgements, either because we or our families are not directly involved, or because we think we are helping society. Now let us consider a situation that will affect us all, either now or in the very near future. For this, instead of working from the general to the particular we need to go from the particular to the general. Let us recap on the issues facing us with regard to genetic testing for disease.

The Institutes of Health in the United States now recommend the routine testing of all couples for cystic fibrosis. This is not mandatory, of course, but neither is it as straightforward in practice as it sounds. The ramifications could be severe, depending upon who thinks they should have access to the results of a test, whether government or medical insurance company. I cannot think that an insurance company would look very favourably on a couple deciding to go ahead with an affected pregnancy, even though it is their absolute right to do so.

In the case of Huntington's chorea, which as we discussed earlier is a late-onset disease, an individual whose parent suffered from this disease faces a dreadful choice. Either they can be tested and risk finding out that they will without doubt suffer the progressive dementia and early death associated with this devastating disorder or they can choose not to be tested and take the future as it comes, with uncertainty. Again we come up against the status of the individual. The immediate response might be that they should be tested for the condition, but who are we to decide on this? If we are not ourselves at risk then we have no right to say that anyone should or should not be tested.

As we all know, life insurance companies ask all manner of questions before they will accept the risk on an individual. They ask about general health, family history and nowadays questions on specific tests such as AIDS tests have started to appear on application forms. Consider the impact of the increasing use of genetic screening on this industry. As the number of genes known to be directly associated with diseases increases and we are more able to detect the mutation responsible, there is growing concern about the application of such information to insurance.

In fact there are surprisingly few genetic diseases that have straightforward single gene patterns of inheritance like cystic fibrosis. Most genetic diseases are associated with what is called polygenic inheritance, that is multiple genes, where several genes influence the onset and severity of a disease, such as hypertension. There are also a number of conditions in which the presence of a gene associated with a disease does not automatically mean that the disease will appear. If you have the gene in this case it does not mean that you will suffer from the disease nor, indeed, does not having the gene mean that you will not become a sufferer. This is contrary to the rather poor understanding of many of the insurance companies.

An example of the unpredictable behaviour of the link between a gene and its associated disease can be found in breast cancer. It is known that breast cancer is associated with a gene called BRCA1. About 5% of breast cancers are the result of this gene, and about 70% of those women who carry the gene will die of the disease before they reach the age of 70. But that leaves 30% of carriers who never succumb to the disease. Exactly what is it that protects this lucky 30%? Nobody knows, and there is no method available to predict if an individual is in the 70% or the 30%.

Given this unpredictability, should insurance companies offering health insurance be allowed access to genetic information which could make some individuals uninsurable, remembering that they may not have any long-term implications for that particular person? This clearly demonstrates the problems associated with over-simple and blanket interpretations of highly complex data. If the pace of genetic research continues it could eventually lead to perfectly healthy people paying grossly inflated premiums because of incorrect assumptions about their genetic background. As a corollary to that, those people who really need insurance may not be able to afford it, while those that do not need it, genetically very healthy, will be the only ones who are able to afford the premiums.

Let us take a closer look at the workings of the insurance industry. There are two basic ways in which insurance and its calculation of premiums work. The first is called mutuality and is the process of keeping all those people of similar risks together and consequently sharing the costs. With mutuality, only people of low risk are welcome onto the books of a company. This is generally the manner in which insurance is calculated in the United States, where the provision of health care and concomitant treatments are almost entirely in the private sector. Private insurance is provided by private companies and this inevitably requires the provision of profit for shareholders in those companies.

In the UK the method is slightly different. It is called solidarity and ostensibly seems a far more equitable system. Basically, everyone shares the costs of insurance no matter what the risks, and unless there are specific modifiers, such as smoking and obesity, generally speaking genetic information can be ignored using this model. If it became commonplace to take into account information from genetic tests, besides the inherent intrusion of an insurance company asking for such private information, which they have a questionable moral right to, they would be doing nothing more than increasing their profits at the expense of human misery.

Think of insurance as a huge game of chance. It has functioned well for the bulk of humanity for many hundreds of years by the construction of actuarial tables which indicate how many people will be expected to die or suffer serious debilitating illness at any given age. There is no indication of who it will be, just the proportion of the general population. On this basis the bet is that the tables will be sufficiently accurate to allow calculation of premiums that will cover all the pay-outs and give the insuring company a profit. To then say that increasing the cost to the potentially affected person is a legitimate business activity is really rather disingenuous. It is merely a method of increasing profits, which have been good enough in past years to produce some of the most profitable companies in the world. Besides which, we live in a complex world and quantifying cause and effect is beyond us at the moment for most of these genetic problems. I would certainly not want an insurance company to introduce a form of eugenics by the back door, which is what would be happening.

Discrimination in any form is a dangerous course to take. We all have something to offer to society, and it is surely neither the purpose nor desire of insurance companies to belittle and cause stress to individuals. It would be quite wrong to label someone as inferior on the basis of spurious and poorly understood data, and this is what may well happen if mutuality becomes the norm. I, like many geneticists, do not believe that the insurance industry has either the skill or the ability to interpret genetic data; it is, after all, not a simple actuarial calculation. It was a sad day in 1998 when the Government of the UK decided to allow insurance companies to use genetic data when assessing risks, especially when this goes directly against the views not only of their own advisory group, the Human Genetics Advisory Commission, but also of the Nuffield Council on Bioethics. Although it would be nice to believe that insurance companies are benign, altruistic groups, they are not; they are companies which will do as much as they can to increase their profits, and self-regulation by the Association of British Insurers is hardly any guarantee of long-term stability.

The ethics of cloning

As I have mentioned earlier, one of the biggest upsets to people's sensibilities about genetics in recent years was the cloning of the sheep called Dolly. The announcement of the success of this experiment stirred up a hotbed of debate and we were bombarded with a staggering number of misinformed ideas on the subject. The general level of ignorance of the media, the public, politicians and religious leaders on the science and practice of genetic research was plain to see. Sadly, as in many issues of public concern, those least equipped to comment had the greatest propensity to do so.

A very good example of this was the level of debate on the question of cloning from a somatic cell. Almost invariably the responses turned to the advisability of cloning a human – as though this had never been done. Nobody seemed willing or able to point out that every pair of identical twins are clones, that is genetically identical. Admittedly, in the case of Dolly she is identical to her parent instead of to her sister and the process was carried out artificially rather than naturally, but does this really matter? Remember that humans have been reproducing themselves, with an occasional clone, since our first appearance as an extant species and we do it very well. Although I must admit that it is a constant surprise to me that so many normal people arise out of such an odd and quirky process, there does not seem any reason to alter the system, it works very well and people enjoy it too.

There is no dispute that a human genetic clone generated from a cell taken from an adult would result in a different individual to the donor. It is inconceivable that two people separated by several years and raised under different circumstances, even if they were genetically identical, would end up the same. To take an extreme example, if we were able to clone Hitler today, it is clear that 50 years after his death, upbringing and environment would not result in the same outcome or the same person. Similarly the idea of cloning a prematurely deceased loved one would be a mistake. The clone would be quite a different individual, not merely many years younger, but also modified from the original person, by who knows how much, simply due to the changes in upbringing. It is probable that they would look the same, as do identical twins, but in personality and behaviour they would be totally different.

There are a number of questions which should be asked about DNA technology and its ethical use, but it is both premature, and frankly unlikely, that cloning of a human from a somatic cell will take place. Having discussed the reasons for pursuing the goal of animal cloning, it should be evident why human cloning is neither desirable nor likely. Not least of all because the one question which we have not addressed but which would be paramount to a clone is the question of identity. One of the universally recognized human rights is the right to a name and hence an identity. We all closely guard our identity from our name to our signature, and cloning would clearly compromise these cornerstones of our individuality.

In the case of plants, it is possible to produce clones quite simply. The regeneration of entire plants from single cells is well documented, as is taking cuttings and producing a complete, genetically identical plant to the parent donor of the cutting. Although the genetic manipulations that have been carried out on plants do not seem to generate quite the same level of shock as those carried out on animals, this is probably only due to familiarity with the process. The production of genetically identical individuals, either plants or animals, has many potential advantages, but also pitfalls which have to be addressed.

One of the main reasons for cloning animals and plants is the production of a profusion of identical embryos from a single individual plant or animal that displays some feature of advantage. This could be a result of genetic manipulation or could be what in older usage was called a 'sport' or a 'rogue', a naturally occurring variation. In most cases, sports are unsuitable for cultivation, but consider what would happen if one of these spontaneously mutated individuals showed some particularly fine and desirable characteristics. The initial response would be to try and breed more and more of the particular plant or animal displaying the desired trait. But there is, inevitably, a catch in all this. What if the newly produced variety, bred as a homogenous crop, whether plant or animal, were to be susceptible to a particular disease or pathogen?

This, of course, is the crux of the practical problem of cloning. When the high-yielding crop is doing well, we all do well. Food is produced in quantities which could not have been otherwise produced. But again we have an anomaly: the production of large quantities of food is fine, so long as it is in the right place for its consumption. Often it is not, often it is in entirely the wrong place and the overproduction is maintained as a simple political or commercial expedient. Many ethical considerations come into play here and I do not want to delve too deeply into the issues surrounding transnational corporations and global agricultural policies.

In the final analysis, it is productivity and disease and pest resistance that have been the most sought after goals of the plant and animal breeders from time immemorial. The introduction of new molecular biology techniques has made it easier to produce this result, but with genetically identical individuals the risks are also increased.

If one single individual succumbs to a disease or parasite, with no variation in the crop, the results would be catastrophic. This is one of the fundamental dangers with cloning. But cloning also gives the possibility of new, high-yield strains with a particular ability to grow in hitherto inhospitable conditions, sometimes without the use, or excessive use anyway, of fertilizers and pesticides. Some people would say that the end result – an increased supply of food on our overcrowded planet – justify the risks involved.

Finally, to return to the issue of human cloning, the question remains: why would we want to clone people? We can already manipulate our environment so as to be able to survive in the depths of the oceans, the vacuum of space and all points in between. We do not need to clone ourselves for reasons of survival, especially as the overproduction of humans is already a well-established fact.

Besides this, we do not entirely rely upon our natural immunity to keep us fit and well, although we do not necessarily rely on natural catastrophes to kill and maim us either. Basically, we do quite well enough thank you without the attendant risks of reducing our genetic diversity at our own hands.

DNA profiling

Another highly contentious area of DNA use, and one which has the potential to burst in upon our lives at any time, is DNA profiling. This is not a simple case of diagnosis and treatment, it can come upon any of us at any time and has the potential to destroy not only your own life but your family's life as well. An adverse result means you will lose your job, and if you are not very lucky, your friends as well. What is this devastation that can sneak upon any one of us? It is DNA profiling as used in criminal cases. We have dealt with the technology of personal identification in an earlier chapter, so here we will concentrate on the social and ethical implications.

The future of the use of DNA for personal identification is particularly interesting, as we already know much about the direction in which it will proceed in the future. This information comes from the forensic scientists who use the current techniques and who will help determine the direction in which the technology will go.

Does anybody have the right to store DNA data about an individual, even if it is of the most rudimentary kind, as the current DNA profiling information most certainly is? In order to attempt an answer, we need to look at the way in which DNA data are stored. There are two types of DNA data storage – those with names and those without. The first database is an anonymous collection of data which is designed to show how frequencies of specific sequences found in different populations vary. This is used for comparison purposes, making it possible to calculate the likelihood of finding any given profile by accident.

The second type of database contains the profiles of specified individuals with their names. This kind of information is more or less confined to individuals who have been convicted of a criminal offence. I say more or less because it is quite common for some professions to have blood taken and stored just in case it should be needed in the future for DNA analysis. This is the procedure being adopted by the armed forces in the United States and probably many other countries as well. The gruesome truth is that, unlike the familiar images of war we see on film, in real armed conflicts bodies can easily be blown up or burnt to unrecognizable pieces. It may only be through comparison with stored samples of DNA that such remains can be unequivocally identified. This may seem, and most probably is, reasonable to us now, but think on.

Our society, taken from the point of one of the first accurately recordable events, is not very old. This event was the battle between the Medes and the Lydians on 28 May 585 BC. Unlike other points in early history, we can date this battle precisely because we know that it was called off when a solar eclipse darkened the sky. As we can calculate such eclipses retrospectively and there was not another one for many years before or later, we can date the battle with complete certainty. That puts us about 2,500 years away from the event, which is nothing in terms of the age of the planet. But just think how many civilizations have grown, flowered and died in that time, along with their benefactors, tyrants and despots. Consider further this century, and just how many people have died in wars and how many leaders of nations have behaved reprehensibly towards their own and other's people, all in the space of that short period of mud and marble which we think of as a lifetime. So how can we be sure that DNA data stored in good faith now will not be misused in the future? Society as a whole does not have a good long-term record of holding true to promises and responsibilities of this kind.

It is not just straightforward DNA data about an individual that has potential for misuse, although that is bad enough, it is the accumulation of information regarding associations of characters together which could also be misused. This particular idea is not as silly as it might at first sound. In this respect the DNA database held by the UK Home Office seems destined to run before it can walk; it represents a clear case of the accumulation of knowledge without concomitant accumulation of responsibility. They may, quite inadvertently, produce information that could be used for quite different purposes in the future. It has been said by a spokesperson from the UK Forensic Science Service that further work on DNA for forensic uses will include determination of specific traits. Stated boldly like that we can see that there could be reasons why this would be useful in criminal investigations. If, for example, a DNA sample could be used to say that a hooded rapist was blond and had green eyes, it would make the work of the police much easier. They could simply remove the majority of the male population from the equation. But this sort of genetic analysis is fraught with danger for society. Assuming it is possible to determine such things, and there are good reasons why they are not easy, let alone possible, testing any such suspect groups may be socially divisive. For example, the technicians involved in this may inadvertently diagnose either a propensity to, or a direct diagnosis of, a genetic disease. Now this is something which the Forensic Science Service has stated categorically must be avoided, but like the insurance companies they probably do not have the expertise to interpret such complicated data correctly. Unfortunately, geneticists know that assertions of certainty about inadvertently revealing a genetic lesion is just not possible. It is virtually impossible to be sure that metabolic pathways associated with specific polygenic traits, like hair colour, are not also linked to, or even directly responsible for, a disease condition.

Hair colour is quite a good feature to consider for this. We know, through genetic studies in mice, that there are probably about 50 genes which control hair colour in humans. This means that any analysis of hair colour would be extremely complex and just may not be worth the effort, certainly not in the short term. And this does not take into account the tricky question of accidentally determining defective genes in an individual and all that may result in.

So where does that leave us when we are deciding whether or not it is a good idea to take and store DNA samples for the identification of dead soldiers? You either accept the consequences or you question them. If after questioning them you find it is acceptable then this is your society and you live in it. But, and it is a big but here, if you do not, then what are you going to do about it?

Morals, ethics and personal freedom – the philosophical view

Since we have looked across a range of possibilities on a piecemeal basis it would be useful to look at how philosophers approach the subject of morals, ethics and personal freedom in the context of

DNA data. From the outset one thing is plain, philosophical considerations are always aimed at making generalities which can then be applied to specific cases. This is similar in a way to the attempts of physics to generate a unified theory of matter. We arrive at simple rules of wide-ranging power, framed so that they can be applied to all manner of situations.

One of the first questions we need to ask is whether there is a place in ethical and moral judgements for emotive arguments and responses. If we accept there is space for emotive reasoning, this is not inconsistent with, and neither does it rule out, using logical arguments to change an attitude or standpoint. This use of both logical and emotive points in an argument assumes that all protagonists are rational, but in one very easily understood and familiar situation this is simply not correct. I refer here to propaganda, which, by definition, is unquestioningly partisan. Taking such emotive arguments out of an equation moves us towards a state where a reason for, or against, an action in one case must be equally for or against it in all similar cases. Put another way and taken on a personal level, if I judge an action to be right for me, then it must be right for all others in a similar situation.

It may never be possible to remove the emotive content from practical arguments involving genetics since they have the potential to impinge not only on all of our lives, but those of future generations also. Under such circumstances it may not even be desirable to do so, but it must be desirable to be consistent in thought and action or we have a state of intellectual anarchy. Witness the shooting in America by an anti-abortionist of a gynaecologist who carried out abortions. If you are against abortion then you have some, presumably, reasoned idea of the sanctity of life. So how can that be consistent with murdering another sentient being? In philosophy this is referred to as a self-excepting idea. Put simply, a self-excepting idea is one that says it is all right for me to do something, but it is not for others. In this case, it is all right for me to murder but not for others.

We would, I think, prefer what was called by Samuel Clarke in 1705 the Rule of Equity. This is something which we all recognize and it cuts across all societies and cultures. In fact it is so universal in historical terms that it may be a lynch-pin in these moral debates. Clarke was not the first propounder of this notion, but being a philosopher, he was the first to label the idea with a name. Basically what he said was that what I judge reasonable or unreasonable that another should do for me is also reasonable or unreasonable that I should do for him. It was not, of course, an original thought, but had been of significance in forming our society for many centuries. If we move forward a little, in October 1747 the Earl of Chesterfield wrote in a letter to his son, 'Do as you would be done by is the surest method that I know of pleasing.' Although that is a little later than Clarke we can still see where his phrase originated; it came from our translation of what can be simply described as religious beliefs or, as some would prefer, moral imperatives.

In the Book of Common Prayer, instituted in 1549 by Edward VI of England, the Catechism includes the phrase 'To love Him as myself, and to do to all men, as I would they should do unto me.' A rather more stern injunction is found in the Bible itself. The King James version of 1611 captures the idea in the sentence: 'Therefore all things whatsoever ye would that men should do to you, do ye even so unto them: for this is the law and the prophets.' Then again, the rather more direct 'thou shalt love thy neighbour as thyself' appears in both the Old and New Testaments.

Just in case you are tempted to imagine that this idea is a purely western Judeo-Christian thought, it also appears as 'What I do not want others to do to me, I have no desire to do to others' in the Analects of Confucius, which was compiled by his pupils after his death in 479 BC.

We can see from this that social cohesion has often been associated with an institutionalized form of altruism; not that governments necessarily take any heed of this, they merely expect their people to do so! Moral points of view tend to exert themselves in this direction because of our natural tendency to ask people to 'put yourself in my shoes'. This naturally results in the question 'Would I like to be on the receiving end of what I am advocating?' Immanuel Kant held that we can only believe in the rational basis of our morality if there is a teleological basis for it. Well, times have changed radically since the eighteenth century, and we no longer need to believe in a theistic being to justify our moral position.

Theorists on the Moral Point of View try very hard to set up systems that can be used almost like an equation to solve difficult problems, but I, like many others, suspect that these generalities can only give guidance. They are unlikely to render an exact technique in which all problems are both amenable to analysis and answerable.

Moral rights with regards to the individual do not always fare well in philosophy. Often glossed over, moral rights may well be quintessential in the discussion of a situation where DNA and genetics are involved. The distinguished philosopher David Hume, who also worked in the eighteenth century, epitomized this when he suggested that the commonsense belief in the obligation of promises can be explained but not justified. Such hard-nosed attitudes can hardly be accommodated in today's world where very few people can be said to be completely self-sufficient. We all depend not only on each other but also on people's promises, even if this is only in some form of trade, for some part of our material survival. As well as this we increasingly all depend and rely on the promises of government that they will not exercise their power in the use of weapons of mass destruction. We are all, in this respect as well, dependent on others and their promises, but on a scale which could never be imagined by philosophers of past centuries.

When we try to force a fit between moral and ethical theory and the questions raised by genetics, it is not always easy to do so. If, on the other hand, we accept a more ancient and, perhaps, more fundamental attitude towards ethical considerations of genetics we can see a light at the end of the tunnel, even though we might not necessarily like the conclusions we reach.

If we do not want anybody to do anything we would not do to ourselves, we are not so much taking a moral standpoint as reflecting our own self-interest. At the same time, is there any point at which we can say that we have a right to make someone do something they do not want to, just because we would? Both of these questions have to be addressed when the termination of an affected pregnancy is involved. But there is another question raised by genetics, which I think should also be considered. This goes back to the all too frequent disregard of individuals when morals are considered. What right has an individual, group or government to question the decisions of another?

As society becomes more homogeneous and information about individuals becomes common currency amongst those that should, and sometimes should not have it, personal decisions become ever more important to the individual. Quite rightly so. Many of the questions arising from DNA data are very personal, so personal in fact that no matter what the philosophical, political or social aspects are, perhaps no one has any right to the information except the individual to whom it refers. No proselytizing terrorist, officially sanctioned or not, should ever hold sway over the public, whether with bomb and bullet or some spurious notion of genetic fitness. And remember that while we respect the laws that say 'no spitting' because they are for the common comfort of society, when the law says 'genetic testing before breeding' that impoverishes us all by reducing our genetic diversity and puts an intolerable burden on the individual and perhaps, in the not too distant future society, on you and me by reducing the long-term survival of *Homo sapiens*.

Appendix

How to extract DNA in your kitchen

Anyone who has ever worked in science knows that laboratories are little more than very well-equipped kitchens. Admittedly this is usually said tongue-in-cheek, but it is not too far from the truth, as this recipe shows. Although this recipe will not produce quite the same performance as a laboratory technique, either in volume or most certainly in purity, this is only because the domestic products used here are not as pure as the undiluted chemicals used in the lab.

If you find that your attempts to produce DNA do not work first time, try reducing the quantities. The commonest mistakes are too much starting material, in which case the result is a sludge, or too much vinegar, water and alcohol, in which case the precipitated DNA is far too dilute to remove. The first time I extracted DNA the result was a dreadful mess because I had started off with so much material, blood in that particular case, that it clotted and was unworkable.

Like all recipes, this one is open to interpretation. You may find that some things do not work well for you and that you need to adapt the technique in some way. Scientific recipes are very often based on a skill unknown to the person exercising it. So if the recipe does not work first time, just persevere; very few of us have got it right first time. On seeing my frustration at a method that did not seem to work, an eminent professor once said to me that when he was in that position he went back to something he knew he could do, even if it was an irrelevant process. This would gradually clear the air until he could tackle the new experiment with confidence. This is very sound advice and I can recommend it.

You will need . . .

First of all, it would be a good idea to read the recipe right through before starting, to make sure everything is at hand for use. Then you can gather together the ingredients:

1. Something containing DNA in large quantities. This can be animal or vegetable, but I would recommend plant material. Animal material, such as liver, can give a higher yield, but it does tend to require additional steps which are not easily carried out in a kitchen. Plant material is much easier to handle and will give just as good a result. Having said that, choose material that is not green. Chorophyll, the green substance in leaves, is not always easy to get rid of. Onion, especially if actively growing, or wheat germ are excellent starting materials. Avoid large-celled watery material like citrus fruits. These have too small a DNA content per gram to make them worth using.
2. Household salt. Not large grained sea salt as this can take time to dissolve, but ordinary table salt.
3. Vinegar. Any sort will do, but clear distilled malt vinegar is best.
4. Washing-up liquid. This is best if it is in the form of a simple strong detergent.
5. Ethanol. Luckily the alcohol in drinks is ethanol. Don't be tempted to use methylated spirits, this is quite different chemically. What you will need is the strongest vodka you can get, certainly nothing less than 37.5% alcohol by volume, which is standard strength for clear spirits. For this experiment some needs to be at room temperature, but most needs to be kept in a freezer until it is very cold.

By way of equipment all you will need is a bowl, a measuring jug and spoons, weighing scales and something to strain the bits out of the solution.

What you have to do

1. First you need to get rid of all the protein of the cell, or as much as possible. This is done by mixing about 100g of your chosen starting material with 150ml of water containing about ½ teaspoon of salt and 1 teaspoon of vinegar. This should be stirred for about 15 minutes while the cell walls are disrupted.
2. Gently strain the material. This can be done using a tea strainer, a paper coffee filter or a muslin jelly bag, as used in jam making. It is the pulp which is required, so do not throw it away.
3. Mix 150ml of cold water with a tablespoon of alcohol, ½ a teaspoon of salt and ½ a teaspoon of washing-up liquid.
4. Add the pulp to the mixture and stir gently for half an hour. During this time the DNA is being dissolved into the liquid.
5. Add a dessertspoon of salt to the solution and stir for another 10 minutes.
6. Leave the solution to stand until it clears, as the solids settle to the bottom.
7. The DNA is in the liquid part, so keep that and discard the solids.
8. Dilute the liquid with three times its volume of ice-cold vodka. Gently stir the mixture as you add it and you should see the DNA precipitate out as white threads. Leave them to settle and there you have it – DNA extracted in the kitchen.

What you do with it is up to you!

Index

Page numbers in *italic* refer to illustrations